Lecture Notes in Mathematics Vol. 334

ISBN 978-3-540-47010-6 © Springer-Verlag Berlin Heidelberg 2008

Fritz Schweiger

The Metrical Theory of Jacobi-Perron Algorithm

Errata

It was not before 1997 that a serious flaw in the proof of theorem 8.7. was detected by Adriana Berechet. An attempt to save the result can be found in F. Schweiger: Kuzmin's theorem revisited. *Ergodic Theory and Dynamical Systems* 20 (2000), 557–565. However, the new proof gives a weaker convergence rate.

More information on the Jacobi-Perron algorithm and other multidimensional continued fractions can be found in A. J. Brentjes: *Multi-dimensional continued fraction algorithms. Mathematical Centre Tracts* 145, Amsterdam: Mathematisch Centrum 1981 and in F. Schweiger: *Multidimensional Continued Fractions.* Oxford: Oxford University Press 2000.

Lecture Notes in Mathematics

A collection of informal reports and seminars
Edited by A. Dold, Heidelberg and B. Eckmann, Zürich

334

Fritz Schweiger
Universität Salzburg, Salzburg/Österreich

The Metrical Theory of Jacobi-Perron Algorithm

Springer-Verlag
Berlin · Heidelberg · New York 1973

AMS Subject Classifications (1970): 10-02, 10A30, 10F10, 10F20, 10K10, 10K15, 10K99, 28A10, 28A70, 28A65

ISBN 3-540-06388-9 Springer-Verlag Berlin · Heidelberg · New York
ISBN 0-387-06388-9 Springer- Verlag New York · Heidelberg · Berlin

Offsetdruck: Julius Beltz, Hemsbach/Bergstr.

Preface

In these Lecture Notes a valuable monograph on the algebraic and arithmetic aspects of Jacobi algorithm by L. Bernstein [1] appeared. This book should be a counterpart to it. It covers almost all aspects of Jacobi algorithm not touched in the beautiful work by L. Bernstein. These are questions of measure theory, ergodic theory, dimension and diophantine approximation. The algorithm is treated as a model of an f-expansion (see Rényi [1]) where all difficulties of multidimensionality enter. There are included some general results on ergodic theory and dimension theory.

I want to thank my teacher W.M. Schmidt to whom I owe my introduction to Jacobi algorithm and to L. Bernstein who persuaded me to write these notes.

My thanks also go to Mrs. Millonig for her patient job of typing and to Dr. Fischer who discovered a lot of errors in previous versions.

University of Salzburg
March 1973

F. Schweiger

Table of Contents

The metrical theory of Jacobi - Perron algorithm

Notation

$[\Theta]$	is the integral part of Θ
B	n-dimensional unit cube after a suitable set of Lebesgue measure zero has been removed
λ	n-dimensional Lebesgue measure
T	the basic transformation as defined in § 1
μ	the basic invariant measure (§ 6)
N	the natural numbers

§ 1. Basic definitions

General references are Perron [1] and Bernstein [1].
We begin with some formal definitions. For n fixed we define the following set of (n+1)x(n+1) matrices

$$\Lambda_o = \begin{pmatrix} 0 & 0 & \dots & 0 & 1 \\ 1 & 0 & \dots & 0 & 0 \\ 0 & 1 & \dots & 0 & 0 \\ \dots & & & & \\ 0 & 0 & \dots & 1 & 0 \end{pmatrix}$$

$$\Lambda_g = \begin{pmatrix} 0 & 0 & \dots & 0 & 1 \\ 1 & 0 & \dots & 0 & a_{g1} \\ \dots & & & & \\ \dots & & & & \\ 0 & 0 & \dots & 1 & a_{gn} \end{pmatrix} \qquad g = 1,2,\dots$$

where $a_g = (a_{g1},\dots,a_{gn})$ is an integral vector. One sees

$$\det \Lambda_g = (-1)^n$$

Furthermore we define

$$\Omega_g = \Lambda_o \Lambda_1 \cdots \Lambda_{g-1}, \quad g \geq 1$$

$$\Omega_o = E \qquad \text{(unit matrix)}$$

Then we have

$$\Omega_{g+1} = \Omega_g \Lambda_g \qquad \text{for all } g \geq 0$$

Denoting $\Omega_g = ((A_i^{(g+j)}))$, $i,j = 0,\ldots,n$

we see

$$A_i^{(j)} = \delta_{ij} \qquad \text{(Kronecker delta)},\ i,j = 0,\ldots,n$$

The definition of the $A_i^{(g+j)}$ is easy to be seen well posed due to the special nature of Λ_g , more precisely

$$\underbrace{A_i^{(g+1+j-1)}}_{\text{as an element of } \Omega_{g+1}} = \underbrace{A_i^{(g+j)}}_{\text{as an element of } \Omega_g} \qquad \text{for } j = 1,\ldots,n$$

and

$$A_i^{(n+1)} = 0 \qquad 1 \leq i \leq n$$

$$A_o^{(n+1)} = 1$$

and

$$A_i^{(g+n+1)} = A_i^{(g)} + \sum_{j=1}^{n} A_i^{(g+j)} a_{gj}$$

for $0 \leq i \leq n$, $g \geq 1$.

We note

$$\det \Omega_g = \det ((A_i^{(g+j)})) = (-1)^{ng}$$

We now set

$$K = \{x \mid 0 \leq x_i < 1\}.$$

and

$$W = \{x \mid x_1 = 0\}$$

Then we define the following mapping

$$T : K \setminus W \to K$$

$$T(x_1,\ldots,x_n) = \left(\frac{x_2}{x_1} - \left[\frac{x_2}{x_1}\right], \ldots, \frac{1}{x_1} - \left[\frac{1}{x_1}\right]\right)$$

Then T^s is defined on

$$K \setminus \bigcup_{j=0}^{s-1} T^{-j}W \quad \text{recursively by the formulae}$$

$$T^o = 1_K$$

$$T^{j+1} = T\,T^j \qquad 0 \leq j \leq s-1$$

We now define

$$B = K \quad \bigcup_{j=0}^{\infty} T^{-j}\,W$$

Then $T : B \rightarrow B$
will be basic for all further developments. Note that T^s is defined on B for all $s \geq 0$. Next we define the following sequence of functions

$$k_s : B \rightarrow N_o^n$$

$$x \rightarrow k_s(x)$$

where N_o^n denotes the set of all vectors with nonnegative integral components:

$$k_1(x) = \left(\left[\frac{x_2}{x_1}\right],\ldots,\left[\frac{1}{x_1}\right]\right)$$

$$k_s(x) = k_1(T^{s-1}\,x) \qquad s \geq 1$$

For $x \in B$ the vector $k_s(x) = (k_{s1}(x),\ldots,k_{sn}(x))$ is in fact a vector in N_o^n. Associating to $k_s(x)$ the matrix

$$A_s(x) = \begin{pmatrix} 0 & \ldots & 0 & 1 \\ 1 & \ldots & 0 & k_{s1} \\ \ldots & \ldots & \ldots & \ldots \\ 0 & \ldots & 1 & k_{sn} \end{pmatrix}$$

we can define for each x the functions $A_i^{(s+j)}(x)$ as explained at the beginning of this chapter.

<u>Lemma 1.1:</u> $\bigcup_{j=0}^{\infty} T^{-j}\,W$ is a countable union of countable pieces of hyperplanes.

<u>Proof:</u> W is a piece of the hyperplane $x_1 = 0$. It is enough to show that the intersection of a hyperplane E with K has a counterimage consisting of countable pieces of hyperplanes.

$$\alpha_1 Y_1 + \alpha_2 Y_2 + \ldots + \alpha_n Y_n + \alpha_o = 0$$

Each $y = (y_1,\dots,y_n) \in K$ has the at most countable counterimages

$T^{-1}y = x = \left(\frac{1}{m_n+y_n}, \dots, \frac{m_{n-1}+y_{n-1}}{m_n + y_n}\right)$ for $m = (m_1,\dots,m_n)$ with

$0 \le m_i \le m_n$ and $1 \le m_n$ as can be seen from the definition of T (in case that $y_{i-1} \ge y_n$, $i = 2,\dots,n$, only $m_i < m_n$ is allowed). Substituting this we have

$$\alpha_1\left(\frac{x_2}{x_1} - m_1\right) + \alpha_2\left(\frac{x_3}{x_1} - m_2\right) + \dots + \alpha_n\left(\frac{1}{x_1} - m_n\right) + \alpha_0 = 0$$

This gives the equation of a hyperplane and $T^{-1}E$ consists of the intersections of these hyperplanes with the regions

$$\{m_1 \le \frac{x_2}{x_1} < m_1+1, \dots, m_n \le \frac{1}{x_1} < m_n+1\} \cap K$$

<u>Lemma 1.2:</u> If $T^s x = y$, $x \in B$, the following relations hold:

$$x_i = \frac{A_i^{(s+n+1)} + \sum_{j=1}^{n} A_i^{(s+j)} y_j}{A_0^{(s+n+1)} + \sum_{j=1}^{n} A_0^{(s+j)} y_j}$$

<u>Proof:</u> By induction. $s = 0$.

$$x_i = \frac{A_i^{(n+1)} + \sum_{j=1}^{n} A_i^{(j)} y_j}{A_0^{(n+1)} + \sum_{j=1}^{n} A_0^{(j)} y_j} = y_i$$

by the definitions of the $A_i^{(j)}$. We assume the formula proved for s and will show its truth for s+1. We put:

$T^s x = y$, $T^{s+1} x = Ty = z$.

The definition of T gives:

$$y_j = \frac{k_{s+1,j-1} + z_{j-1}}{k_{s+1,n} + z_n} \qquad \text{for } 2 \le j \le n$$

$$y_1 = \frac{1}{k_{s+1,n} + z_n}$$

From this we have $(0 \le g \le n)$

$$A_g^{(s+n+1)} + \sum_{j=1}^{n} A_g^{(s+j)} y_j = A_g^{(s+n+1)} + \sum_{j=2}^{n} A_g^{(s+j)} \frac{k_{s+1,j-1} + z_{j-1}}{k_{s+1,n} + z_n} +$$

$$+ A_g^{(s+1)} \frac{1}{k_{s+1,n} + z_n} = \frac{1}{k_{s+1,n} + z_n} (A_g^{(s+n+1)} k_{s+1,n} +$$

$$+ \sum_{j-1=1}^{n-1} A_g^{(s+1+j-1)} k_{s+1,\, j-1} + A_g^{(s+1)} + A_g^{(s+n+1)} z_n +$$

$$+ \sum_{j-1=1}^{n-1} A_g^{(s+1+j-1)} z_{j-1}) = \frac{1}{k_{s+1,n} + z_n} (A_g^{(s+1+n+1)} + \sum_{j=1}^{n} A_g^{(s+1+j)} z_j)$$

Substituting this in the formula for s we get our result.

We now give the following definition:

A sequence $\omega = (a_1, a_2, \ldots)$ of vectors $a_i \in N_o^n$ will be called admissible if the following conditions hold:

(i) $0 \leq a_{si} \leq a_{sn}$, $i = 1,\ldots,n$, $s \geq 1$

(ii) $1 \leq a_{sn}$

(iii) The relations $(0 \leq t \leq i-1)$

$a_{si} = a_{sn}$

$a_{s+1,i-1} = a_{s+1,n-1}$

$\ldots\ldots\ldots\ldots\ldots\ldots$

$a_{s+t,i-t} = a_{s+t,n-t}$

imply

$a_{s+t+1,i-(t+1)} \leq a_{s+t+1,n-(t+1)}$

if $t+1 < i$, and

$1 \leq a_{s+t+1,n-(t+1)}$

if $t+1 = i$.

Lemma 1.3: If the sequence $(a_1, a_2, \ldots)$ is admissible, the sequence $(a_2, a_3, \ldots)$ is admissible.

Proof: Clear.

Lemma 1.4: Given an admissible sequence $(b_1, b_2, b_3, \ldots)$ the sequence

$(a_1,\ldots,a_g,b_1,b_2,\ldots)$, $g \geq n-1$, is admissible if and only if

(α) $(a_1,\ldots,a_g)$ is admissible that means, is the beginning of at least one admissible sequence

(β) $(a_{g-n+2},\ldots,a_g,b_1,b_2,\ldots)$ is admissible.

Proof: The "only if" part is clear by Lemma 1.3.

The conditions (i) and (ii) are satisfied for $(a_1,\ldots,a_g,b_1,b_2,\ldots)$ by (α). The condition (iii) has only influence to at most the next n-1 vectors (as can be seen by the worst case $i = n-1$ and $t = n-2$).

Lemma 1.5: For any $x \in B$ the sequence $(k_1(x),\ k_2(x),\ldots)$ is admissible.

Proof: It is enough to prove the relations for $s = 1$.

(i) $0 \leq x_{i+1} < 1,\ x_1 > 0$ gives

$$0 \leq \frac{x_{i+1}}{x_1} < \frac{1}{x_1} \qquad 1 \leq i \leq n-1$$

Hence $0 \leq [k_{1i}] \leq [k_{1n}]$

(ii) Clear

(iii) $$x_{i+1} = \frac{k_{1i} + (Tx)_i}{k_{1n} + (Tx)_n} \qquad 1 \leq i \leq n-1$$

If $k_{1i} = k_{1n}$ from $0 \leq x_{i+1} < 1$ we have

$$(Tx)_i < (Tx)_n$$

and hence

$$k_{2,i-1} \leq k_{2,n-1}$$

Using

$$(Tx)_i = \frac{k_{2,i-1} + (T^2x)_{i-1}}{k_{2n} + (T^2x)_n}$$

$$(Tx)_n = \frac{k_{2,n-1} + (T^2x)_{n-1}}{k_{2n} + (T^2x)_n}$$

we have if $k_{2,i-1} = k_{2,n-1}$ is valid too

$$(T^2x)_{i-1} < (T^2x)_{n-1}$$

and hence

$$k_{3,i-2} \leq k_{3,n-2}$$

Concluding in a similar fashion we get the result.

We now define X the set of all admissible sequences and a mapping

$$T^{*} : X \to X$$

$$T(a_1, a_2, \ldots) = (a_2, a_3, \ldots)$$

Furthermore we define

$$\Phi : B \to X$$

$$\Phi(x) = (k_1(x), k_2(x), \ldots)$$

The following result is easy:

Lemma 1.6:

$$T^{*}\Phi = \Phi T$$

Not easy to prove is the following

Theorem 1.7 (Perron [1]):

The mapping Φ is bijective.

Sketch of proof:

1.Step: $\omega = (a_1, a_2, \ldots)$ be any admissible sequence. To ω we associate the functions $A_i^{(g+j)}$ via

$$A_g = \begin{pmatrix} 0 & \ldots\ldots & 1 \\ 1 & \ldots\ldots & a_{g1} \\ \ldots & \ldots & \ldots \\ 0 & \ldots\ldots & a_{gn} \end{pmatrix}$$

Then we define the sequences

$$\left(\frac{A_i^{(n+s)}}{A_o^{(n+s)}} \right) \qquad s = 1, 2, \ldots \qquad \text{for } i = 1, \ldots, n$$

It will be shown the sequences converge. We put

$$x_i = \lim_{s \to \infty} \frac{A_i^{(n+s)}}{A_o^{(n+s)}}$$

and $x = (x_1, \ldots, x_n)$. Then we have a mapping

$$\Psi : X \to B$$

$$\Psi(\omega) = \mathbf{x}$$

Naturally one must show $x \in B$.

<u>2.Step:</u> $\Psi\,\Phi = 1_B$

This means: Given an $x \in B$ the sequence $\Phi(x)$ is admissible. Forming the $A_i^{(s+j)}$, $i = 0,\dots,n$ we have the convergence

$$\lim_{s\to\infty} \frac{A_i^{(n+s)}}{A_o^{(n+s)}} = x_i$$

<u>3.Step:</u> $\Phi\,\Psi = 1_X$

This means: Given an admissible sequence ω we can associate the point $x \in B$ via $x = \Psi(\omega)$. Then $k_s(x) = a_s$.

§ 2. Cylinders

Let $a_1,\dots,a_m$ be any sequence of integral n-dimensional vectors.

We define:

$$B(a_1,\dots,a_m) = \{x \in B \mid k_1(T^{s-1}x) = a_s,\ 1 \le s \le m\}$$

$B(a_1,\dots,a_m)$ is called a cylinder of order m.

<u>Lemma 2.1:</u> $B(a_1,\dots,a_m) \neq \emptyset$ iff $a_1,\dots,a_m$ is the beginning of an admissible sequence.

<u>Proof:</u> Clear.

Henceforth we will only consider cylinders defined by admissible sequences.

We list some properties obvious from the definition

(1) $B(a_1,\dots,a_m) \cap B(b_1,\dots,b_m) = \emptyset$ for $(a_1,\dots,a_m) \neq (b_1,\dots,b_m)$

(2) $B(a_1,\dots,a_m) = \bigcup B(a_1,\dots,a_m, t)$ where t runs over all integral vectors for which $a_1,\dots,a_m, t$ is admissible

(3) $\lambda(B(a_1,\dots,a_m)) = \sum_t \lambda(B(a_1,\dots,a_m, t))$

(4) $\sum_{a_1,\dots,a_m} \lambda(B(a_1,\dots,a_m)) = \lambda(B) = 1$

The last equality is by lemma 1.1.

(5) B itself is the cylinder of order 0.

<u>Lemma 2.2:</u> $T^s B(a_1,\dots,a_s,a_{s+1},\dots,a_{s+t}) = B(a_{s+1},\dots,a_{s+t})$ for all $t \ge n-1$ and $s \ge 1$.

Proof: The admissibility of a sequence $b_1, b_2, b_3, \dots$ to give an admissible sequence $a_1, \dots, a_{s+t}, b_1, b_2, \dots$ depends only on the n-1 predecessors:

$$a_{s+t-(n-2)}, a_{s+t-(n-3)}, \dots, a_{s+t}$$

We now give the following definition:

A cylinder $B(a_1, \dots, a_m)$ will be called proper if $T^m B(a_1, \dots, a_m) = B$. In general only $T^m B(a_1, \dots, a_m) \subseteq B$ is valid. Later on we will study the partition generated by the $T^m B(a_1, \dots, a_m)$ more carefully. At present we only use:

Lemma 2.3:

$M = \{x \mid 0 < x_1 < x_2 < \dots < x_n\} \subsetneq T^m B(a_1, \dots, a_m)$

Proof: From lemma 2.2. we may assume $m = n-1$. In fact for $m \geq n-1$ we have $T^m B(a_1, \dots, a_m) = T^{n-1} T^{m-n+1} B(a_1, \dots, a_m) = T^{n-1} B(a_{m-n+2}, \dots, a_m)$

Inspection of the proof of lemma 1.4 shows that the restrictions of digits yield relations of the form

$$t_i < t_j$$

for $t = (t_1, \dots, t_n)$. The worst case gives $0 \leq t_1 < t_2 < t_3 < \dots < t_n < 1$

A more subtile geometric description of the $B(a_1, \dots, a_m)$ which are in fact intersections of convex polytopes with B can be found in F.Schweiger [4]. A lot of geometric interpretation was done by R.Meijer [1] in an unpublished work. From this we see that

$$T^s : B(a_1, \dots, a_s) \rightarrow B$$

is a mapping from the cylinder $B(a_1, \dots, a_s)$ on a certain subset of B which contains M. We denote this subset with $P(a_1, \dots, a_s)$. The restriction of T^s to $B(a_1, \dots, a_s)$ has an inverse function

$$V(a_1, \dots, a_s) : P(a_1, \dots, a_s) \rightarrow B(a_1, \dots, a_s)$$

Basic for the whole theory will be

Lemma 2.4: $V(a_1, \dots, a_s)$ is differentiable on $P(a_1, \dots, a_s)$ and the absolute value of its functional determinant $\Delta(a_1, \dots, a_s)$ is given by

$$\Delta(a_1, \dots, a_s)(y) = \frac{1}{\left(A_0^{(s+n+1)} + \sum_{j=1}^{n} A_0^{(s+j)} y_j\right)^{n+1}}$$

$(y \in P(a_1, \dots, a_s))$

Proof: With $T^s x = y$ we have

$$x_i = \frac{A_i^{(s+n+1)} + \sum_{j=1}^{n} A_i^{(s+j)} y_j}{A_o^{(s+n+1)} + \sum_{j=1}^{n} A_o^{(s+j)} y_j} \qquad \text{for } x \in B(a_1,\dots,a_s).$$

Hence $V(a_1,\dots,a_s)$ is the restriction of an fractional linear transformation to $P(a_1,\dots,a_s)$. Partial differentiation gives

$$\frac{\partial x_i}{\partial y_k} = \frac{A_i^{(s+k)} (A_o^{(s+n+1)} + \sum_{j=1}^{n} A_o^{(s+j)} y_j)}{(A_o^{(s+n+1)} + \sum_{j=1}^{n} A_o^{(s+j)} y_j)^2} -$$

$$- \frac{A_o^{(s+k)} (A_i^{(s+n+1)} + \sum_{j=1}^{n} A_i^{(s+j)} y_j)}{(A_o^{(s+n+1)} + \sum_{j=1}^{n} A_o^{(s+j)} y_j)^2} = \frac{A_i^{(s+k)} - x_i A_o^{(s+k)}}{A_o^{(s+n+1)} + \sum_{j=1}^{n} A_o^{(s+j)} y_j}$$

We have to compute the determinant of the n x n - matrix

$$((A_i^{(s+k)} - x_i A_o^{(s+k)})) \quad , \quad 1 \le i, \quad k \le n$$

This determinant is equal to the larger determinant:

$$\begin{vmatrix} 1 & A_o^{(s+1)} & A_o^{(s+2)} & \dots & A_o^{(s+n)} \\ 0 & A_1^{(s+1)} - x_1 A_o^{(s+1)} & A_1^{(s+2)} - x_1 A_o^{(s+2)} & \dots & A_1^{(s+n)} - x_1 A_o^{(s+n)} \\ \vdots & \vdots & \vdots & & \vdots \\ 0 & A_n^{(s+1)} - x_n A_o^{(s+1)} & A_n^{(s+2)} - x_n A_o^{(s+n)} & \dots & A_n^{(s+n)} - x_n A_o^{(s+n)} \end{vmatrix} =$$

$$\begin{vmatrix} 1 & A_o^{(s+1)} & A_o^{(s+2)} & \dots & A_o^{(s+n)} \\ x_1 & A_1^{(s+1)} & A_1^{(s+2)} & \dots & A_1^{(s+n)} \\ \vdots & \vdots & \vdots & & \vdots \\ x_n & A_n^{(s+1)} & A_n^{(s+2)} & \dots & A_n^{(s+n)} \end{vmatrix} =$$

$$= \frac{1}{A_o^{(s+n+1)} + \sum_{j=1}^{n} A_o^{(s+j)} y_j} \qquad \det\left((A_i^{(s+j)}) \right) \quad 0 \leq i\,, \quad j \leq n$$

The last equality follows from lemma 1.2.

Corollary 2.5:

$$\sup_{y\varepsilon P(a_1,\ldots,a_s)} \Delta(a_1,\ldots,a_s)\ (y) \leq C \inf_{y\varepsilon P(a_1,\ldots,a_s)} \Delta(a_1,\ldots,a_s)(y)$$

With $C = (n+1)^{n+1}$

Proof:

$$\sup \Delta(a_1,\ldots,a_s) \leq \frac{1}{(A_o^{(s+n+1)})^{n+1}}$$

$$\inf \Delta(a_1,\ldots,a_s) \geq \frac{1}{(A_o^{(s+n+1)} + \sum_{j=1}^{n} A_o^{(s+j)})^{n+1}}$$

$$\geq \frac{1}{(A_o^{(s+n+1)})^{n+1} \quad (1 + \sum_{j=1}^{n} A_o^{(s+j)} \ / \ A_o^{(s+n+1)})^{n+1}}$$

$$\geq \frac{1}{(A_o^{(s+n+1)})^{n+1} \quad (n+1)^{n+1}}$$

Lemma 2.6:

$$\frac{\sup \Delta(a_1,\ldots,a_s)}{n! \quad (n+1)^{n+1}} \leq \lambda(B(a_1,\ldots,a_s)) \leq (n+1)^{n+1} \inf \Delta(a_1,\ldots,a_s)$$

Proof:

$$\lambda(B(a_1,\ldots,a_s)) \leq \int_B \Delta(a_1,\ldots,a_s)\ d\lambda \leq$$

$$\leq \sup \Delta(a_1,\ldots,a_s) \leq (n+1)^{n+1} \inf \Delta(a_1,\ldots,a_s)$$

$$\lambda(B(a_1,\ldots,a_s)) \geq \int_{P(a_1,\ldots,a_s)} \Delta(a_1,\ldots,a_s) d\lambda \geq$$

$$\geq \inf \Delta(a_1,\ldots,a_s)(n!)^{-1} \geq$$

$$\geq \sup \Delta(a_1,\ldots,a_s) \ \left[n! \quad (n+1)^{n+1}\right]^{-1}$$

We note the general formula

$$\lambda(B(a_1,\ldots,a_s)) = \int_{P(a_1,..,a_s)} \Delta(a_1,\ldots,a_s)d\lambda$$

Lemma 2.7:

$$\Delta(a_1,\ldots,a_s,\ a_{s+1},\ldots,a_{s+t})(y) =$$

$$= \Delta(a_1,\ldots,a_s)(V(a_{s+1},\ldots,a_{s+t})y)\Delta(a_{s+1},\ldots,a_{s+t})(y)$$

Proof: This follows directly from the product rule for Jacobians. Note that the admissibility of the sequence $a_1,\ldots,a_s$, $a_{s+1},\ldots,a_{s+t}$ is reflected by

$$B(a_{s+1}),\ldots,a_{s+t}) \cap P(a_1,\ldots,a_s) \neq \emptyset$$

Corollary 2.8:

$$c_1\lambda(B(a_1,\ldots,a_s))\ \lambda(B(a_{s+1},\ldots,a_{s+t})) \leq \lambda(B(a_1,\ldots,a_{s+t})) \leq$$

$$\leq c_2\lambda(B(a_1,\ldots,a_s))\ \lambda(B(a_{s+1},\ldots,a_{s+t}))$$

Proof: Note that

$$\lambda(B(a_1,\ldots,a_s)) \leq \sup \Delta(a_1,\ldots,a_s)$$

and (using lemma 2.3)

$$\lambda(B(a_1,\ldots,a_s)) \geq (n!)^{-1} \inf \Delta(a_1,\ldots,a_s).$$

We now denote for a metric space (X,d) with

$$\operatorname{diam} M = \sup \{d(x,y) \mid x \in M\ ,\ y \in M\}$$

the diameter of a subset M.

We set

$\sigma(m) = \sup \operatorname{diam} B(a_1,\ldots,a_m)$ where the supremum runs over all admissible sequences $a_1,\ldots,a_m$.

The metric space in question is B as a subspace of the unit cube K with the Euclidean metric. In fact it can be seen easily that the diameter of a cylinder remains unaltered if one takes the cylinder a subset of W instead of B(a cylinder $B(a_1,\ldots,a_m)$ is an intersection of a nondegenerate convex polytope with 2^n faces with B and B is taken from W by omission of countably many pieces of hyperplanes).

Clearly $\sigma(m)$ is a decreasing function. This follows from

$$B(a_1,\ldots,a_{m+1}) \subseteq B(a_1,\ldots,a_m)$$

Basic is the following

Lemma 2.9: $\lim_{m\to\infty} \sigma(m) = 0$

Proof: We first show

$$\lim_{s\to\infty} \operatorname{diam} B(k_1,\dots,k_s) = 0$$

for every admissible sequence $k_1,k_2,\dots$. Given an admissible sequence every $x \in B(k_1,\dots,k_s)$ may be written as

$$x = \sum_{j=1}^{n+1} f_j^{(s)} p^{(s+j)}$$

$$0 < f_j^{(s)} \leq 1$$

$$\sum_{j=1}^{n+1} f_j^{(s)} = 1$$

$$p^{(s+j)} = \left(\frac{A_1^{(s+j)}}{A_0^{(s+j)}}, \dots\dots, \frac{A_n^{(s+j)}}{A_0^{(s+j)}}\right)$$

$$f_j^{(s)} = \frac{A_0^{(s+j)} \; y_j}{\sum_{j=1}^{n+1} A_0^{(s+j)} \; y_j}$$

$(y = T^s x,\ y_{n+1} = 1$ formally).

By the main convergence theorem we see

$$\lim_{s\to\infty} p^{(s+j)} = z = \Psi(k_1,k_2,\dots.)$$

Hence we have

$$d(x,z) \leq \sum_{j=1}^{n+1} d(p^{(s+j)}, z) < \frac{\varepsilon}{2} \quad \text{if } s \geq s(\varepsilon).$$

Therefore $\operatorname{diam} B(k_1,\dots,k_s) < \varepsilon$ for $s \geq s(\varepsilon)$.

We will now show the following: For every $\delta > 0$ and $s \geq 1$ there are only finitely many cylinders $B(k_1,\dots,k_s)$ with

$$\operatorname{diam} B(k_1,\dots,k_s) \geq \delta$$

We again use

$$x_i = \frac{\sum_{j=1}^{n+1} A_i^{(s+j)} \; y_j}{\sum_{j=1}^{n+1} A_o^{(s+j)} \; y_j}, \qquad y = T^s x$$

$$x = (x_1,\ldots,x_n) \in B(k_1,\ldots,k_s)$$

This gives

$$\sum_{j=1}^{n+1} y_j (x_1 A_o^{(s+j)} - A_i^{(s+j)}) = 0$$

and

$$x_i - \frac{A_i^{(s+n+1)}}{A_o^{(s+n+1)}} = - \sum_{j=1}^{n} \frac{y_j A_o^{(s+j)}}{A_o^{(s+n+1)}} \left(x_i - \frac{A_i^{(s+j)}}{A_o^{(s+j)}}\right)$$

Using $0 \leq y_j \leq 1$ for $1 \leq j \leq n$ we get

$$\left| x_i - \frac{A_i^{(s+n+1)}}{A_o^{(s+n+1)}} \right| \leq \sum_{j=1}^{n} \frac{A_o^{(s+j)}}{A_o^{(s+n+1)}} \left| x_i - \frac{A_i^{(s+j)}}{A_o^{(s+j)}} \right|$$

Since clearly $x \in B(k_1,\ldots,k_{s-n+j-1})$ for $1 \leq j \leq n$ we obtain

$$\mathrm{diam}(B(k_1,\ldots,k_s)) \leq 2 \sum_{j=1}^{n} \frac{A_o^{(s+j)}}{A_o^{(s+n+1)}} \cdot \mathrm{diam}(B(k_1,\ldots,k_{s-n+j-1}))$$

This relation holds for $s \geq n$. The initial relations $(0 < s < n)$ are obtained formally by deleting the superfluous summands with $s-n+j-1<0$. From

$$A_o^{(s+n+1)} \geq k_{s,n} \cdot \ldots \cdot k_{s-n+j,n} \; A_o^{(s+j)} \quad \text{we see by induction:}$$

Given n cylinders $B(k_1,\ldots,k_{s-n+j-1})$, $1 \leq j \leq n$, there are only finitely many $k_{s,n}$ with diam $B(k_1,\ldots,k_s) \geq \delta$. Since there are only finitely many k_s for a given $k_{s,n}$, there are only finitely many cylinder

$B(k_1,\dots,k_s)$ with diam $B(k_1,\dots,k_s) \geq \delta$. From this we can infer that in fact

$$\sigma(m) = \max \operatorname{diam} B(a_1,\dots,a_m)$$

Now we can finish the proof. Suppose that $\sigma(m) \geq \eta > 0$ for all $m \geq 1$. This implies: For every $m \geq 1$ there is a sequence $k_1,\dots,k_m$ with diam $B(k_1,\dots,k_m) \geq \eta$. Since for fixed m the set of these sequences is finite there must be a least one a_1 with the property:For every $m \geq 1$ there is a sequence $a_1,k_2,\dots,k_m$ with diam $B(a_1,k_2,\dots,k_m) \geq \eta$. (Here we used the obvious fact $B(k_1,\dots,k_{m-1}) \supseteq B(k_1,\dots,k_m)$).

Repeating this argument we find an a_2 with the property: For every $m \geq 2$ there is a sequence $a_1,a_2,k_3,\dots,k_m$ with diam $B(a_1,a_2,k_3,\dots,k_m) > \eta$. Hence we find a sequence $a_1,a_2,a_3,\dots$ with diam $B(a_1,\dots,a_s) \geq \eta$ for all $s \geq 1$. This contradicts $\lim\limits_{s\to\infty} \operatorname{diam} B(a_1,\dots,a_s) = 0$.

This proof can be possibly generalized to a broader class of multidimensional f-expansions. In fact an assertion like lemma 2.9 is implicitly used in Waterman's papers on Kuzmin's theorem ([1], [2]). Recently Fischer [1] gave a new short proof for lemma 2.9. Put

$$\rho = \left(1 - \frac{1}{(n+1)^n}\right)^{1/n}$$

then $\sigma(s) = O(\rho^s)$. A sketch of the proof will be presented in § 9.

§ 3. Increasing σ-fields

We define as usual:

Def.: A collection $\mathcal{F}$ of subsets of a set Ω is called a σ-field if the following conditions hold:

(i) If $(A_i)_{i\in N}$ is a countable family of sets $A_i \in \mathcal{F}$, then

$$\bigcup_{i\in N} A_i \in \mathcal{F}$$

(ii) For any $A \in \mathcal{F}$ it follows $\Omega \setminus A \in \mathcal{F}$. A set $A \in \mathcal{F}$ often is called measurable. From the definition follows at once:

(iii) $\emptyset \in \mathcal{F}$, $\Omega \in \mathcal{F}$

(iv) For a countable family $(A_i)_{i\in N}$ with $A_i \in \mathcal{F}$ also $\bigcap\limits_{i\in N} A_i \in \mathcal{F}$.

Proposition 3.1. For any collection $(\mathcal{F}_\alpha)_{\alpha\in A}$ of σ-fields (over Ω), $\bigcap\limits_{\alpha\in A} \mathcal{F}_\alpha$ is a σ-field.

Proof: (i) $(M_i)_{i\in\mathbb{N}}$ be a countable collection, $M_i \in \bigcap_{\alpha\in A} \mathcal{F}_\alpha$. Then $M_i \in \mathcal{F}_\alpha$ for all $\alpha\in A$ and $\bigcup_{i\in N} M_i \in \mathcal{F}_\alpha$ for all $\alpha\in A$.

(ii) For $M \in \bigcap_{\alpha\in A} \mathcal{F}_\alpha$ clearly $\Omega \setminus M \in \bigcap_{\alpha\in A} \mathcal{F}_\alpha$.

Corollary 3.2. If $\mathcal{E}_\alpha$, $\alpha\in A$, are collections of subsets of Ω, there exists exactly one smallest σ-field containing all sets of $\bigcup_{\alpha\in A} \mathcal{E}_\alpha$. This σ-field will be denoted with $\bigvee_{\alpha A} \mathcal{E}_\alpha$.

Proof: The collection P of all σ-fields $\mathcal{F}$ containing all sets of $\bigcup_{\alpha\in A} \mathcal{E}_\alpha$. is not empty (the set of all subsets of Ω is a σ-field contained in P). Clearly

$$\bigvee_{\alpha\in A} \mathcal{E}_\alpha = \bigcap_{\mathcal{F}\in P} \mathcal{F}$$

$\bigvee_{\alpha\in A} \mathcal{E}_\alpha$ is called generated by $(\mathcal{E}_\alpha)_{\alpha\in A}$

Examples:

(1) $\mathcal{Z}^{(s)}$ be the set of all cylinders $B^{(s)} = B(a_1,\ldots,a_s)$ of order s.

We denote

$\mathcal{L}^{(s)} = \vee\, \mathcal{Z}^{(s)}$ where $\vee\, \mathcal{E}$ is the σ-field generated by $\mathcal{E}$.

$\mathcal{L}^{(s)}$ is the smallest σ-field containing all cylinders of order $\leq$ s. Its members are easily seen countable unions of cylinders of order s. (Note that every cylinder of order $t < s$ is a countable union of cylinders of order s)

We remark the obvious facts

(1) $\mathcal{L}^{(0)} = \{\emptyset, B\}$

(2) $\mathcal{L}^{(s)} \subseteq \mathcal{L}^{(s+1)}$ for all $s \geq 0$

(2) For any metric space (X,d) the collection of Borel sets $\mathcal{B}$ is the σ-field generated by the open (or closed) sets. In the sequel we are most interested in the Borel field $\mathcal{B}$ of the metric space B in its induced natural metric (given by Euclidean distance).

Note that from a Lebesqueian point of view the Borel field $\mathcal{B}$ of B is not very different from the Borel field $\mathcal{B}^*$ of K (see lemma 1.1).

Lemma 3.3. $\bigvee_{s\in N} \mathcal{Z}^{(s)} = \mathcal{B}$

<u>Proof:</u> From the fact that every cylinder $B^{(s)}$ is Borelian we have
$\bigvee_{s\in N} \mathfrak{Z}^{(s)} \subseteq \mathcal{L}$

Given a closed set F, let M_s be the union of all cylinders of $\mathfrak{Z}^{(s)}$ that intersect F. Clearly $F \subseteq M_s$. From lemma 2.9 we have

$$F = \bigcap_{s\in N} M_s$$

because $\sigma(s) \downarrow 0$ and every point of M_s has its distance from F less than $\sigma(s)$. Therefore $\bigvee_{s\in N} \mathfrak{Z}^{(s)}$ contains all closed sets and

$$\mathcal{L} \subseteq \bigvee_{s\in N} \mathfrak{Z}^{(s)}$$

Given a sequence of σ-fields

$$\mathcal{G}^{(1)} \subseteq \mathcal{G}^{(2)} \subseteq \mathcal{G}^{(3)} \subseteq \ldots \quad \text{with} \quad \bigvee_{s\in N} \mathcal{G}^{(s)} = \mathcal{G}$$

we write $\mathcal{G}^{(s)} \uparrow \mathcal{G}$ and call this sequence an increasing sequence of σ-fields with limit $\mathcal{G}$. Lemma 3.3. may be restated in the form

$$\ell^{(s)} \uparrow \mathcal{L}$$

<u>§ 4. Conditional expectations</u>

<u>Def.</u>: A triple $(\Omega,\mathcal{F},P)$ is called a probability space if

(1) Ω is a set

(2) $\mathcal{F}$ is a σ-field of subsets of Ω

(3) P is a probability measure

The last condition means that P is a real valued function on $\mathcal{F}$ satisfying

(3.1) $A \subseteq B$ implies $P(A) \leq P(B)$

(3.2) If $(A_n)_{n\in N}$ is a sequence of pairwise disjoint measurable sets we have

$$\sum_{n\in N} P(A_n) = P(\bigcup_{n\in N} A_n)$$

(3.3) $P(\emptyset) = 0$

(3.4) $P(\Omega) = 1$

Examples will be the space $(B,\mathcal{L},\lambda)$ and the spaces $(B,\ell^{(s)},\lambda)$. In these cases λ is the measure induced by n-dimensional Lebesgue measure.

We recall some definitions known from the theory of integration and measure:

Def.: A real-valued function $f : \Omega \to \mathbb{R}$ on a space $(\Omega, \mathcal{F}, P)$ is called measurable with respect to $\mathcal{F}$ if $f^{-1} A \in \mathcal{F}$ for all Borel sets of the real line.

Def.: A function $\tau : \mathcal{F} \to \mathbb{R}$ on a σ-field $\mathcal{F}$ is called σ-additive if

$$\sum_{n \in N} \tau(A_n) = \tau(\bigcup A_n)$$

for any sequence of pairwise disjoint sets $A_n \in \mathcal{F}$ and $\tau(\emptyset) = 0$.

Any probability measure is σ-additive but not conversely.

Def: Given two σ-additive functions

$$\tau_1, \tau_2 : \mathcal{F} \to \mathbb{R}$$

we call τ_1 absolutely continuous on $\mathcal{F}$ with respect to τ_2 if

$$\tau_2(F) = 0 \text{ implies } \tau_1(F) = 0.$$

For any integrable function $f : \Omega \to \mathbb{R}$ on $(\Omega, \mathcal{F}, P)$ and any σ-field $\mathcal{G} \subseteq \mathcal{F}$ we define

$$\tau : \mathcal{G} \to \mathbb{R}$$
$$\tau(G) = \int_G f$$

(the integral taken with respect to the measure P). Clearly τ is a (finite) σ-additive function. From $\mathcal{G} \subseteq \mathcal{F}$ we have τ absolutely continuous on $\mathcal{G}$ with respect to P. This means:

$$\tau(G) = 0 \text{ for every set } G \in \mathcal{G} \text{ with } P(G) = 0.$$

The Radon - Nikodym theorem (see Halmos [2] or Munroe [1] p.196) gives the existence of a function $E(f \| \mathcal{G})$ which satisfies

(α) $E(f \| \mathcal{G})$ is integrable on $(\Omega, \mathcal{G}, P)$ and hence measurable (with respect to $\mathcal{G}$)

(β) $\int_G E(f \| \mathcal{G}) = \int_G f$

Def.: A function $E(f \| \mathcal{G})$ obeying (α) and (β) is called a conditional expectation of f with respect to $\mathcal{G}$.

In general there are many conditional expectations but they all agree with the exception of a set $G \in \mathcal{G}$ with $P(G) = 0$ (abbreviated: almost everywhere = a.e.).

We give two examples:

① $\mathcal{G} = \mathcal{F}$ then obviously
$E(f \| \mathcal{F}) = f$ a.e.

(2) Take $\mathcal{G} = \{\emptyset, \Omega\}$ then
$E(f \| \mathcal{G}) = \int_\Omega f$ a.e.

The following results are immediate:

(1) If $f \leq g$, then
$E(f \| \mathcal{G}) \leq E(g \| \mathcal{G})$ a.e.

(2) $E(af + bg \| \mathcal{G}) = aE(f \| \mathcal{G}) + bE(g \| \mathcal{G})$ a.e.
for a and b constants.

(3) $|E(f \| \mathcal{G})| \leq E(|f| \| \mathcal{G})$ a.e.

(4) If f is measurable for $\mathcal{G}$, then
$E(f \| \mathcal{G}) = f$ a.e.

Lemma 4.1: For any integrable function f on $(B, \mathcal{L}, \lambda)$ we may take

$$E(f \| \mathcal{L}^{(s)})(x) = \frac{1}{\lambda(B(a_1, \ldots, a_s))} \int_{B(a_1, \ldots, a_s)} f$$

for $x \in B(a_1, \ldots, a_s)$.

Proof: Conditions (α) and (β) are clearly satisfied.

Lemma 4.2. Let $(\Omega, \mathcal{F}, P)$ be a probability space and $\mathcal{G}^{(s)} \uparrow \mathcal{G}$ an increasing sequence of σ-fields contained in $\mathcal{F}$. For an integrable f and any real number $\lambda > 0$ we can prove

$$P\{\sup_{s \geq 1} E(f \| \mathcal{G}^{(s)}) \geq \lambda\} \leq \frac{1}{\lambda} \int_\Omega |f|$$

Proof: We show

$$P\{\max_{1 \leq t \leq s} E(f \| \mathcal{G}^{(t)}) \geq \lambda\} \leq \frac{1}{\lambda} \int_\Omega |f|$$

We introduce the sets

$$M_t = \{x \mid E(f \| \mathcal{G}^{(u)}) < \lambda \text{ for } u < t \text{ and } E(f \| \mathcal{G}^{(t)}) \geq \lambda\}$$

The sets $M_1, \ldots, M_s$ are pairwise disjoint and their union is the set in question. $E(f \| \mathcal{G}^{(u)})$ is measurable with respect to $\mathcal{G}^{(u)}$ and hence measurable with respect to $\mathcal{G}^{(t)}$ for $u \leq t$ (here the assumption $\mathcal{G}^{(s)} \uparrow$ is used).
Therefore $M_t \in \mathcal{G}^{(t)}$. This gives

$$\lambda P(M) = \sum_{t=1}^{n} \int_{M_t} \lambda \leq \sum_{t=1}^{n} \int_{M_t} E(f \| \mathcal{G}^{(t)}) =$$

$$= \sum_{t=1}^{n} \int_{M_t} f \leq \int_{\Omega} f$$

Theorem 4.3: Under the conditions of lemma 4.2 we have

$$\lim_{s\to\infty} E(f \,\|\, \mathcal{G}^{(s)}) = E(f \,\|\, \mathcal{G}) \qquad \text{a.e.}$$

Note: A.e. means that equality holds with exception of a set $G \in \mathcal{G}$ with $P(G) = 0$.

Proof: We first assume that f is measurable with respect to $\mathcal{G}$. Then $E(f \| \mathcal{G}) = f$ a.e. If f is measurbale with respect to some $\mathcal{G}^{(s)}$ clearly $E(f \| \mathcal{G}^{(r)}) = f$ a.e. for $r \geq s$ and the theorem is obvious.

We note first:
Given a set $G \in \mathcal{G}$ and $\varepsilon > 0$ we can find a set $H \in \mathcal{G}^{(s)}$, $s = s(G,\varepsilon)$, such that

$$P(G \,\Delta\, H) < \varepsilon$$

In the general case f may be approximated by a simple function (see Munroe [1]) measurable with respect to $\mathcal{G}$. From $\mathcal{G}^{(s)} \uparrow \mathcal{G}$ it follows that we may approximate by simple functions measurable with respect to $\mathcal{G}^{(s)}$, more precisely : To $\varepsilon > 0$ we can choose a function f_ε, measurable with respect to $\mathcal{G}^{(s)}$ for $s = s(\varepsilon)$, such that

$$\int_{\Omega} |f - f_\varepsilon| < \varepsilon$$

Hence for m and $n \geq s(\varepsilon)$ we have (using f_ε measurable for $\mathcal{G}^{(m)}$ and $\mathcal{G}^{(n)}$)

$$|E(f \| \mathcal{G}^{(m)}) - E(f \| \mathcal{G}^{(n)})| \leq |E(f_\varepsilon \| \mathcal{G}^{(m)}) - E(f_\varepsilon \| \mathcal{G}^{(n)})| +$$

$$+ |E(f \| \mathcal{G}^{(m)}) - E(f_\varepsilon \| \mathcal{G}^{(m)})| + |E(f \| \mathcal{G}^{(n)}) - E(f_\varepsilon \| \mathcal{G}^{(n)})|$$

$$\leq 2 \sup_{s>1} E(|f - f_\varepsilon| \,\|\, \mathcal{G}^{(s)}) \qquad \text{a.e.}$$

Now $P\{\sup_{s\geq 1} E(|f - f_\varepsilon| \,\|\, \mathcal{G}^{(s)}) \geq \lambda\} \leq \frac{1}{\lambda} \int_{\Omega} |f - f_\varepsilon| \leq \frac{\varepsilon}{\lambda}$ by lemma 4.2

From this we see that the sequence $E(f \| \mathcal{G}^{(s)})$ is fundamental a.e. (with respect to $\mathcal{G}$).

We set $h = \lim_{s\to\infty} E(f \| \mathcal{G}^{(s)})$. Since by Fatou's lemma (see e.g. Munroe [1] p.243)

$$\int |h| \leq \liminf \int |E(f \| \mathcal{G}^{(s)})| \leq \liminf \int E(|f| \,\|\, \mathcal{G}^{(s)}) \leq \int |f|$$

we see h integrable.

Furthermore we see

$$\int |E(f \| \mathcal{G}^{(s)}) - h| \leq \int_{N_\lambda} |E(f \| \mathcal{G}^{(s)}) - h| + \int_{\Omega\setminus N_\lambda} E(|f| \| \mathcal{G}^{(s)}) + \int_{\Omega\setminus N_\lambda} |h|$$

Here $N_\lambda = \{x \mid E(f \| \mathcal{G}^{(s)}) < \lambda\}$. Clearly $N_\lambda \in \mathcal{G}^{(s)}$.

Therefore we have

$$\int |E(f \| \mathcal{G}^{(s)}) - h| \leq \int_{N_\lambda} |E(f \| \mathcal{G}^{(s)}) - h| + \int_{\Omega\setminus N_\lambda} (|f| + |h|)$$

For fixed λ we let $s \to \infty$ and see the first term tend to zero by the Lebesgue theorem on dominated convergence. By lemma 4.2. we have $P(\Omega\setminus N_\lambda) \to 0$ with $\lambda \to \infty$.

Therefore we have proved

$$\lim_{s\to\infty} \int E(f \| \mathcal{G}^{(s)}) = \int h$$

For $A \in \mathcal{G}^{(t)}$ we have

$$\lim_{s\to\infty} \int_A E(f \| \mathcal{G}^{(s)}) = \int_A h \quad \text{and clearly}$$

$$\int_A E(f \| \mathcal{G}^{(s)}) = \int_A f \quad \text{for } s \geq t. \text{ Hence we see}$$

$$\int_A h = \int_A f \qquad \text{for all } A \quad \bigcup_{s\in N} \mathcal{G}^{(s)}.$$

Since $\mathcal{G}^{(s)} \uparrow \mathcal{G}$ this holds for all $A \in \mathcal{G}$ and therefore $h = f$ a.e.

If the function f is not measurable for $\mathcal{G}$, the theorem is proved for $E(f \| \mathcal{G})$, namely

$$\lim_{s\to\infty} E(E(f \| \mathcal{G}) \| \mathcal{G}^{(s)}) = E(f \| \mathcal{G})$$

From $\mathcal{G}^{(s)} \subseteq \mathcal{G}$ we have $E(E(f \| \mathcal{G}) \| \mathcal{G}^{(s)}) = E(f \| \mathcal{G}^{(s)})$ a.e.

This last equality can be seen from

<u>Lemma 4.4.</u> If f is integrable and $\mathcal{G}_1 \subseteq \mathcal{G}_2$, then

$$E(E(f \| \mathcal{G}_2) \| \mathcal{G}_1) = E(f \| \mathcal{G}_1) \qquad \text{a.e.}$$

<u>Proof:</u> We check conditions (α) and (β). Clearly $E(E(f \| \mathcal{G}_2) \| \mathcal{G}_1)$ is measurable for $\mathcal{G}_1$. Furthermore for $A \in \mathcal{G}_1$ we see

$$\int_A E(E(f \| \mathcal{G}_2) \| \mathcal{G}_1) = \int_A E(f \| \mathcal{G}_2) = \int_A f$$

The last equality is from $A \in \mathcal{G}_2$.

§ 5. Ergodicity of the transformation

Given a probability space $(\Omega,\mathcal{F},P)$ we have the following definitions:

Definition: A mapping $\Theta : \Omega \to \Omega$ is called measurable if $\Theta^{-1} F \in \mathcal{F}$ for every measurable set $F \in \mathcal{F}$.

Definition: A measurable mapping $\Theta : \Omega \to \Omega$ is called ergodic if $\Theta^{-1}E=E$ implies $P(E) = 0$ or $P(E) = 1$.

Before proving the main theorem of this section we state

Lemma 5.1: There is an absolute constant $\gamma > 0$ with

$$\Sigma\lambda(B(a_1,\dots,a_m,a_{m+1})) \geq \gamma\lambda(B(a_1,\dots,a_m))$$

where the left sum goes over all proper cylinders contained in $B(a_1,\dots,a_m)$.

Proof: Recall a cylinder $B^{(s)}$ is called proper if $T^s B^{(s)} = B$. From corollary 2.8. we have $\lambda(B(a_1,\dots,a_m,a_{m+1}))\geq c_1.\lambda(B(a_1,\dots,a_m))\lambda(B(a_{m+1}))$

If the digit $a_{m+1} = (a_{m+1,1},\dots,a_{m+1,n})$ satisfies $a_{m+1,n} > a_{m+1,n-1}>\dots>$ $> a_{m+1,1} \geq 1$ we see by lemma 1.3. that $B(a_1,\dots,a_m,a_{m+1})$ is a proper cylinder (there is no restriction to the digits following a_{m+1}).

Clearly $\Sigma\lambda(B(a_{m+1})) = \beta > 0$ where the sum runs over all digits a_{m+1} as described above is a positive constant. Take $\gamma = c_1\beta$.

Theorem 5.2: The transformation $T : B \to B$ is measurable and ergodic on the space $(B,\mathcal{L},\lambda)$.

Proof: From $T^{-1}\,\mathcal{L}^{(s)} = \mathcal{L}^{(s+1)}$ and lemma 3.3. we see T measurable. Now be M a measurable set with $T^{-1}M = M$ and c_M its indicator function. Then the conditional expectations can be given as

$$E(c_M \,\|\, \mathcal{L}^{(s)})(x) = \frac{1}{\lambda(B^{(s)})} \int_{B^{(s)}} c_M \qquad \text{for } x \in B^{(s)} .$$

For a proper cylinder $B^{(s)}$ we see

$$\int_{B(a_1,\dots,a_s)} c_M = \int_B c_M\,\Delta(a_1,\dots,a_s) \geq c_1.\lambda(B(a_1,\dots,a_s))\lambda(M) \quad \text{by}$$

lemma 2.6.

In the general case we have

$$\int_{B(a_1,..,a_s)} c_M \geq \Sigma \int_{B(a_1,..,a_s,a_{s+1})} c_M \geq$$

$$\geq \Sigma\,\lambda(B(a_1,..,a_s,a_{s+1}))(n+1)^{-(n+1)}\lambda(M) \geq \gamma c_1\,\lambda(M)\,\lambda(B(a_1,\dots,a_s))$$

(the sum runs over all proper cylinders $B(a_1,\dots,a_s,a_{s+1})$ $B(a_1,..,a_s)$

and we used lemma 5.1.)

This gives

$$E(c_M \,\|\, \mathcal{L}^{(s)}) \geq \gamma c_1 \ \lambda(M)$$

From the convergence theorem 4.3. we see $c_M = E(c_M \,\|\, \mathcal{B}) \geq \gamma c_1 \ \lambda(M)$ almost everywhere. From $\lambda(M) > 0$ we conclude $c_M = 1$ a.e.and $\lambda(M)=1$. The proof in Schweiger [2] is more complicated in showing $c_M < 1$ a.e. for $\lambda(M) < 1$.

§ 6. Existence of an equivalent invariant measure

Given a probability space $(\Omega,\mathcal{F},P)$ and a mapping $\Theta : \Omega \to \Omega$ we give the following

Def.: Θ preserves the measure P if $P(A) = P(\Theta^{-1} A)$ for $A \in \mathcal{F}$. Clearly, Θ is assumed to be measurable. P is called an invariant measure (with respect to Θ).

In general a mapping Θ will not preserve a given measure P. In fact the transformation $T : B \to B$ does not preserve Lebesgue measure λ.

Def.: Two measures P and M defined on the same σ-field $\mathcal{F}$ are called equivalent if P is absolutely continuous with respect to M and conversely. This means:

$$P(A) = 0 \quad \text{is equivalent to } M(A) = 0$$

One of the most interesting problems of ergodic theory is to solve the following question: Given a space $(\Omega,\mathcal{F},P)$ and a measurable mapping $\Theta : \Omega \to \Omega$, does in this case exist a measure M invariant to Θ and equivalent to P ? We will only consider a small part of this problem and refer the interested reader to N.A.Friedman [1], K.Jacobs [1], [2], S.R.Foguel [1] , P.Halmos [1].

Def.: For measurable Θ we define a sequence of set functions by:

(i) $\quad P_o = P$

(ii) $\quad P_1(A) = P(\Theta^{-1}A)$

(iii) $\quad P_n(A) = P_{n-1}(\Theta^{-1}A)$

Lemma 6.1: P_n is probability measure on $\mathcal{F}$.

Proof: It is enough to show P_1 a probability measure on $\mathcal{F}$.

(3.1) $\quad A \subseteq B$ implies $\Theta^{-1} A \subseteq \Theta^{-1} B$ hence $P_1(A) \leq P(B)$.

(3.2) $\sum_{n\in N} P_1(A_n) = \sum_{n\in N} P(\Theta^{-1}A_n) = P(\bigcup_{n\in N} \Theta^{-1}A_n) = P(\Theta^{-1} \bigcup_{n\in N} A_n) =$

$= P_1(\bigcup_{n\in N} A_n)$

(3.3) $P_1(\emptyset) = 0$

(3.4) $P_1(\Omega) = 1$

Clearly, P is invariant to Θ if and only if $P_n = P$ for all $n \in N$.

Theorem 6.2: If there exist two constants $c_1, c_2 > 0$ with $c_1 P(A) \leq P_n(A) \leq$ $\leq c_2 P(A)$ (uniformly) for all $A \in \mathcal{F}$, then there exists a probability measure M, invariant with respect to Θ and obeying

$$c_1 \, P(A) \leq M(A) \leq c_2 \, P(A)$$

Hence M is equivalent to P.

Proof: Since we use Banach-Mazur limits we have to introduce some preliminary remarks:

The set S of all bounded sequences of real numbers is a normed linear space if one puts

$$\| s \| = \sup_{n\in N} |s_n|$$

for a sequence $s = (s_n)_{n\in N}$. The set S^* of convergent sequences is a subspace of S. The functions

$$p : S \to \mathbb{R}$$
$$p(s) = \limsup s_n$$
$$f : S \to \mathbb{R}$$
$$f(s) = \liminf s_n$$

coincide on the subspace S^* with the continuous linear functional

$$l : S^* \to \mathbb{R}$$
$$l(s) = \lim_{n\to\infty} s_n$$

By the Hahn-Banach theorem (see K.Yosida [1] , p. 1o2-1o4) it follows the existence of a linear functional

$L : S \to \mathbb{R}$ obeying:

(i) $f(s) \leq L(s) \leq p(s)$

(ii) $L(s) = l(s)$ for $s \in S^*$

(iii) $L(s) = L(t)$

where t is the shifted sequence $(t_{n+1}) = (s_n), n \in N)$.

This functional is called a Banach-Mazur limit.

We now define a set function $\pi : \mathcal{F} \to \mathbb{R}$ by

$$\pi(A) = L(s(A))$$

where $s(A)$ denotes the sequence $(P_n(A))_{n \in N}$ for every $A \in \mathcal{F}$.

We show the following properties hold:

(3.1) $A \subseteq B$ implies $\pi(A) \leq \pi(B)$

(3.2^*) If $A \cap B = \emptyset$ we have

$$\pi(A \cup B) = \pi(A) + \pi(B)$$

(3.3) $\pi(\emptyset) = 0$

(3.4) $\pi(\Omega) = 1$

We first prove (3.2^*): If $A \cap B = \emptyset$, then

$$P_n(A \cup B) = P_n(A) + P_n(B)$$

for all $n \in N$. From the linearity of L we see

$$\pi(A \cup B) = \pi(A) + \pi(B).$$

If $A \subseteq B$ we have $B = A \cup (B \setminus A)$, hence

$$\pi(B) = \pi(A) + \pi(B \setminus A)$$

Since $\pi(B \setminus A) \geq 0$ we have (3.1), (3.3) and (3.4) are obvious. We further note:

$$\pi(\Theta^{-1}A) = \pi(A)$$

This comes from (iii). The assumptions of the theorem imply

$$c_1 P(A) \leq \pi(A) \leq c_2 P(A)$$

Hence we see π is nearly a measure lacking only (3.2) of § 4.

Using a method of A.P.Caldéron [1] we proceed in the following way: We define for $A \in \mathcal{F}$ $\quad M(A) = \inf \{ \sum_{n \in N} \pi(A_n) \mid A_n \in \mathcal{F}, A \subseteq \bigcup_{n \in N} A_n \}$

We now check:

(3.1) Let $A \subseteq B$. For $\varepsilon > 0$ we find a sequence $(B_n)_{n \in N}$ with $B \subseteq \bigcup_{n \in N} B_n$ and $\sum_{n \in N} \pi(B_n) \leq M(B) + \varepsilon$. From this and $A \subseteq \bigcup_{n \in N} B_n$ we find

$$M(A) \leq \sum_{n \in N} \pi(B_n) \leq M(B) + \varepsilon$$

From $\varepsilon > 0$ arbitrary we have the result.

(3.2) Let $A = \bigcup_{n \in N} A_n$, where the sets A_n are pairwise disjoint. For $\varepsilon > 0$ we choose $(A_{nk})_{k \in N}$ with

$$A_n \subseteq \bigcup_{k \in N} A_{nk}$$

$$\sum_{k \in N} \pi(A_{nk}) \leq M(A_n) + \varepsilon 2^{-n}$$

Clearly $A \subseteq \bigcup_{n,k \in N} A_{nk}$ and therefore

$$M(A) \leq \sum_{n,k\in N} \pi(A_{nk}) \leq \sum_{n\in N} (M(A_n) + \varepsilon 2^{-n}) \leq \sum_{n\in N} M(A_n) + \varepsilon$$

This implies $M(A) \leq \sum_{n\in N} M(A_n)$.

Next be $C \cap D = \emptyset$ and $(B_n)_{n\in N}$ a measurable covering of $C \cup D$. Clearly $(C \cap B_n)_{n\in N}$ and $(D \cap B_n)_{n\in N}$ are measurable coverings of C and D respectively. Furthermore $\pi(B_n) \geq \pi(B_n \cap C) + \pi(B_n \cap D)$ by the properties of π.

Hence $\sum_{n\in N} \pi(B_n) \geq \sum_{n\in N} \pi(B_n \cap C) + \sum_{n\in N} \pi(B_n \cap D) \geq M(C) + M(D)$

This is valid for every covering $(B_n)_{n\in N}$ hence $M(C \cup D) \geq M(C) + M(D)$. Extending this to n pairwise disjoint sets we see

$$\sum_{i=1}^{n} M(A_i) \leq M(\bigcup_{i=1}^{n} A_i) \leq M(A) \quad \text{and} \quad \sum_{i\in N} M(A_i) \leq M(A)$$

(3.3) and (3.4) are obvious.

From $A \subseteq \bigcup_{n\in N} A_n$ iff $\Theta^{-1}A \subseteq \bigcup_{n\in N} \Theta^{-1} A_n$ we see $M(A) = M(\Theta^{-1}A)$ for $A \in \mathcal{F}$. This means that M is in fact an invariant measure. Clearly $M(A) \leq \pi(A) \leq$ $\leq c_2 P(A)$

On the other side

$$\sum_{i\in N} \pi(A_i) \geq c_1 \sum_{i\in N} P(A_i) \geq c_1 P(A)$$

for each covering $A \subseteq \bigcup_{i N} A_i$ and this proves $M(A) \geq c_1 P(A)$.

Corollary 6.3: If Θ is ergodic with respect to P, then Θ is ergodic with respect to M.

Proof: For $\Theta^{-1}E = E$ implies $P(E) = 0$ or $P(\Omega \setminus E) = 0$. Therefore $M(E) = 0$ or $M(\Omega \setminus E) = 0$.

We are now in position to prove

Theorem 6.4 (Schweiger [3]: There exists a probability measure μ which is preserved by T and for which T is ergodic obeying

$$c_1 \lambda(A) \leq \mu(A) \leq c_2 \lambda(A)$$

for $A \in \mathcal{L}$. In particular μ is equivalent to λ.

Proof: The theorem already follows from

Lemma 6.5: There exist two constants $c_1, c_2 > 0$ for which

$$c_1\lambda_n\,(B(a_1,\dots,a_s)) \le \lambda(B(a_1,\dots,a_s)) \le c_2\lambda_n(B(a_1,\dots,a_s))$$

is true for each cylinder $B(a_1,\dots,a_s)$.

Remark: In fact if lemma 6.5 is proven, we can conclude

$$c_1\lambda_n\,(A) \le \lambda(A) \le c_2\lambda_n(A)$$

using $\mathcal{L}^{(s)} \uparrow \mathcal{L}$ and theorem 6.2 and corollary 6.3.

Proof: Clearly

$$T^{-h}\,B(a_1,\dots,a_m) = \bigcup_{s_1,\dots,s_h} B(s_1,\dots,s_h,a_1,\dots,a_m)$$

where the summation goes over all $s_1,\dots,s_n$ for which $s_1,\dots,s_h,a_1,\dots,a_m$ is an admissible sequence. From corollary 2.8 we get

$$\lambda(T^{-h}B(a_1,\dots,a_m)) \le$$

$$\le n!\,(n+1)^{n+1} \cdot \sum_{s_1,\dots,s_h} \lambda(B(s_1,\dots,s_h))\lambda(B(a_1,\dots,a_m)) \le$$

$$\le n!\,(n+1)^{n+1}\lambda(B(a_1,\dots,a_m))$$

We used $\sum_{s_1,\dots,s_h} \lambda(B(s_1,\dots,s_h)) \le \lambda(T^{-h}B) = 1.$

For the other direction we observe

$$\lambda(T^{-h}B(a_1,\dots,a_m)) \ge c_1 \sum_{s_1,\dots,s_h} \lambda(B(s_1,\dots,s_h)).\lambda(B(a_1,\dots,a_m)) \ge$$

$$\ge c_3 \sum_{s_1,\dots,s_{h-1}} \lambda(B(s_1,\dots,s_{h-1}))\,.\,\sum_{s_h} \lambda(B(s_h))\lambda(B(a_1,\dots,a_m))$$

If $s_h = (s_{h1},\dots,s_{hn})$ satisfies (*) $s_{hn} > s_{h,n-1} > \dots > s_{h1} \ge 1$ the sequence $s_h,a_1,\dots,a_m$ is admissible and on the other hand $s_1,\dots,s_{h-1},s_n$ is admissible for $s_1,\dots,s_{h-1}$ admissible.

Therefore

$$\sum_{s_1,\dots,s_{h-1}} \lambda(B(s_1,\dots,s_{h-1})) = 1$$

for every sequence $s_1,\dots,s_{h-1},s_h$ where s_h satisfies (*).

From lemma 5.1. we get

$$\sum_{s_h} \lambda(B(s_h)) = \beta$$

for these digits s_h and all together shows

$$\lambda(T^{-h}B(a_1,\dots,a_m)) \geq \beta\, c_3\, \lambda(B(a_1,\dots,a_m))$$

§ 7. The ergodic theorem

We begin with an easy preliminary

Lemma 7.1: Let $(\Omega,\mathcal{F},P)$ be a probability space and $\Theta : \Omega \to \Omega$ a measurable mapping. The set J of all sets $A \in \mathcal{F}$ which satisfy $\Theta^{-1} A = A$ is a sub-σ-field of $\mathcal{F}$.

The sets of J are called invariant.

Proof: We check the conditions:

$$\Theta^{-1} \bigcup_{n\in N} A_n = \bigcup_{n\in N} \Theta^{-1} A_n = \bigcup_{n\in N} A_n \in \quad \text{for } A_n \in J.$$

Next we see $\Theta^{-1}(\Omega \setminus A) = \Omega \setminus \Theta^{-1} A$

Ergodicity of Θ can be expressed as $P(I) = 0$ or 1 for $I \in J$.

Lemma 7.2: Θ is ergodic iff $E(f \,\|\, J) = \int_\Omega f$ a.e. (with respect to J) for any integrable function f.

Proof: We first assume

$$E(f \,\|\, J) = \int_\Omega f \qquad \text{a.e.}$$

Take $f = c_I$ the characteristic function of an invariant set. Clearly c_I is measurable J hence

$$E(f \,\|\, J) = c_I \qquad \text{a.e.}$$

Hence $P(I) = c_I$ a.e. This gives the result $P(I) = 0$ or 1. Now assume Θ be ergodic. The set

$$N_\alpha = \{x \mid E(f \,\|\, J)(x) \leq \alpha\}$$

is measurable J for real α hence invariant. From $P(N_\alpha) = 0$ or 1 and respecting $N_\alpha \subseteq N_\beta$ for $\alpha \leq \beta$ we see $E(f \,\|\, J) = \alpha$ for some constant α a.e. The property

$$\int_\Omega E(f \,\|\, J) = \int_\Omega f$$

gives the result.

We note that from $J \subseteq \mathcal{F}$ we have

$$E(f \,\|\, J) = \int_\Omega f \quad \text{a.e. with respect to } \mathcal{F}.$$

We now assume the probability space $(\Omega,\mathcal{F},M)$ equipped with an invariant measure M.

Lemma 7.3: For integrable f we have

(a) $f(x) \geq 0$ a.e. implies $f(\Theta x) \geq 0$ a.e.

(b) $\int_\Omega f(x)dM = \int_\Omega f(\Theta x)dM$

Proof: Let $E = \{x \mid f(x) < 0\}$ then $M(E) = 0$. Therefore $M(\Theta^{-1}E) = 0$ because M is an invariant measure.

Clearly

$$\Theta^{-1} E = \{x \mid \Theta x \in E\} = \{x \mid f(\Theta x) < 0\}$$

The formula $M(\Theta^{-1}A) = M(A)$ gives (b) for an indicator function c_A. Approximating f by simple functions we get (b).

Lemma 7.4: (Maximal Ergodic Theorem):

Let f integrable and

$$E = \{x \mid \sup_{n \in N} \frac{1}{n} \sum_{k=0}^{n-1} f(\Theta^k x) > 0\}$$

Then

$$\int_E f(x)dM \geq 0.$$

Proof: We follow Garsia [1] and Friedman [1]. We put

$$S_o f(x) \equiv 0 \qquad \text{and}$$

$$S_n f(x) = \sum_{k=0}^{n-1} f(\Theta^k x) \qquad \text{for } n \geq 1$$

Define

$$S_n^+ f(x) = \max_{0 \leq t \leq n} S_t f(x)$$

From $S_o f \equiv 0$ we have $S_n^+ f(x) \geq 0$.

We define

$$E_n = \{x \mid S_n^+ f(x) > 0\}$$

Then clearly $E_n \subseteq E_{n+1}$, $n \in N$ and $\bigcup_{n \in N} E_n = E$. It suffices to prove

$$\int_{E_n} f(x)dM \geq 0$$

We have $S_n^+ f(x) \geq S_t f(x)$ for $0 \leq t \leq n$ and using lemma 7.3 (a) we see

$$S_n^+ f(\Theta x) \geq S_t f(\Theta x), \; 0 \leq t \leq n$$

Clearly $f(x) + S_t f(\Theta x) = S_{t+1} f(x)$ and therefore

$$f(x) + S_n^+ f(\Theta x) \geq S_t f(x) \quad \text{for } 1 \leq t \leq n+1.$$ This implies

$$f(x) + S_n^+ f(\Theta x) \geq \max_{1\leq t\leq n+1} S_t f(x) \quad \text{and for } x \in E_n$$

we see

$$f(x) + S_n^+ f(\Theta x) \geq S_{n+1}^+ f(x) \geq S_n^+ f(x)$$

Integration over E_n implies

$$\int_{E_n} f(x) dM \geq \int_{E_n} (S_n^+ f(x) - S_n^+ f(\Theta x)) dM$$

Since $S_n^+ f(x) = 0$ and $S_n^+ f(\Theta x) \geq 0$ for $x \in \Omega \setminus E_n$ we see

$$\int_{E_n} (S_n^+ f(x) - S_n^+ f(\Theta x)) dM \geq \int_\Omega (S_n^+ f(x) - S_n^+ f(\Theta x)) dM = 0$$

by lemma 7.3 (b).

Theorem 7.5 (Individual or Pointwise Ergodic Theorem): Let $(\Omega, \mathcal{F}, M)$ be a probability space and M invariant with respect to Θ. For every integrable f we have

$$\lim_{n\to\infty} \frac{1}{n} \sum_{k=0}^{n-1} f(\Theta^k x) = E(f \| J) \qquad \text{a.e.}$$

In particular if Θ is ergodic, we get

$$\lim_{n\to\infty} \frac{1}{n} \sum_{k=0}^{n-1} f(\Theta^k x) = \int_\Omega f(x) dM \qquad \text{a.e.}$$

Proof: The last assertion is by lemma 7.2. For $a < b$ put

$$Y = Y(a,b) = \{x \mid \liminf_{n\to\infty} \frac{1}{n} \sum_{k=0}^{n-1} f(\Theta^k x) < a < b <$$

$$< \limsup_{n\to\infty} \frac{1}{n} \sum_{k=0}^{n-1} f(\Theta^k x)\}$$

We easily see Y is an invariant set and therefore $c_Y(\Theta x) = c_Y(x)$ for its indicator function. We apply lemma 7.4 to the function $(f(x) - b)\, c_Y(x)$. Clearly

$$Y \subseteq \{x \mid \sup_{n\in N} \frac{1}{n} \sum_{k=0}^{n-1} f(\Theta^k x) > b\} \quad \text{and therefore}$$

$$\int_Y (f(x) - b) dM \geq 0$$

In the same manner we get

$$\int_Y (a - f(x))dM \geq 0$$

Adding gives

$$\int_Y (a - b)dM = M(Y(a,b))(a - b) \geq 0$$

This implies $M(Y(a,b)) = 0$. If $W = \cup Y(a,b)$, the union over all pairs of rational numbers a,b, we see $M(W) = 0$ and hence we may conclude

$$\lim_{n\to\infty} \frac{1}{n} \sum_{k=0}^{n-1} f(\Theta^k x) = \hat{f}(x) \qquad \text{a.e.}$$

where $\hat{f}(x)$ is finite, $+\infty$ or $-\infty$. By Fatou's lemma

$$\int_\Omega |\hat{f}(x)|\, dM \leq \liminf_{n\to\infty} \int_\Omega \left|\frac{1}{n} \sum_{k=0}^{n-1} f(\Theta^k x)\right| dM \leq$$

$$\leq \liminf_{n\to\infty} \int_\Omega \frac{1}{n} \sum_{k=0}^{n-1} |f(\Theta^k x)|\, dM = \int_\Omega |f(x)|\, dM$$

The last equality is by lemma 7.3.

Thus $\hat{f}$ is integrable and finite-valued a.e.

From

$$\frac{1}{n+1} \sum_{k=0}^{n} f(\Theta^k x) = \frac{f(x)}{n+1} + \frac{n}{n+1} \sum_{k=0}^{n-1} f(\Theta^{k+1} x)$$

we see $\hat{f}(\Theta x) = \hat{f}(x)$ a.e. Hence $\hat{f}$ is measurable J. Now let $I \in J$ be an invariant set. We set

$$Y = Y(a,b) = \{x \in I \mid a \leq \hat{f}(x) < b\}$$

and c_Y be its indicator function.

For $\varepsilon > 0$ we apply again lemma 7.4 to the function $[f - (a - \varepsilon)]\, c_Y$.

Then $Y \subseteq \{x \mid \sup_{n\in N} \frac{1}{n} \sum_{k=0}^{n-1} f(\Theta^k x) > a - \varepsilon\}$ and $\int_Y f(x)dM \geq (a - \varepsilon)M(Y)$.

Since $\varepsilon > 0$ is arbitrary, we obtain

$$\int_Y f(x)dM \geq aM(Y)$$

Similarly we can show

$$bM(Y) \geq \int_Y f(x)dM$$

For fixed n, we put

$$Y_k = Y\left(\frac{k}{2^n}, \frac{k+1}{2^n}\right), \quad k = 0, \pm 1, \pm 2, \ldots \qquad \text{and obtain}$$

$$\frac{k}{2^n} M(Y_k) \leq \int_{Y_k} f(x)dM \leq \frac{k+1}{2^n} M(Y_k)$$

Since this relation is valid for $\hat{f}$ too we see

$$-\frac{1}{2^n} M(Y_k) \leq \int_{Y_k} f(x)dM - \int_{Y_k} \hat{f}(x)dM \leq \frac{1}{2^n} M(Y_k)$$

We sum over all $k = 0, \pm 1, \pm 2, \ldots$ and observe I the disjoint union of all Y_k. We obtain

$$\left| \int_I f(x)dM - \int_I \hat{f}(x)dM \right| \leq \frac{1}{2^n} M(I)$$

This gives

$$\int_I f(x)dM = \int_I \hat{f}(x)dM \qquad \text{and}$$

$$\hat{f} = E(f \| J) \qquad \text{a.e.}$$

We are now in position to obtain a sharpening of theorem 6.2 and corollary 6.3.

Theorem 7.6: $(\Omega,\mathcal{F},P)$ be a probability space and the measurable mapping be ergodic with respect to P. If there exists an invariant measure M equivalent to P it is uniquely determined and

$$M(A) = \lim_{n\to\infty} \frac{1}{n} \sum_{k=0}^{n-1} P_k(A)$$

Proof: Clearly

$$P_k(A) = \int_\Omega c_A (\Theta^k x)dP$$

If there is an invariant measure we have

$$\lim_{n\to\infty} \frac{1}{n} \sum_{k=0}^{n-1} c_A(\Theta^k x) = E(c_A \| J)(x) \qquad \text{a.e.}$$

Since M is ergodic (by corollary 6.3) we have

$$\lim_{n\to\infty} \frac{1}{n} \sum_{k=0}^{n-1} c_A(\Theta^k x) = M(A) \qquad \text{a.e.}$$

This shows M uniquely determined, because M was an arbitrary invariant measure equivalent to P. We used that a.e. with respect to M is a.e. with respect to P and conversely. The Lebesgue dominated convergence theorem ($c_A \leq 1$) yields the asserted formula.

Applying our results to the Jacobi algorithm we get:

Theorem 7.7: There exists an unique measure, equivalent to Lebesgue measure λ and invariant under T.

In the sequel this measure will be denoted with μ.

The ergodic theorem 7.5 allows a number of applications from which we give only several typical:

Theorem 7.8: Let $B(s_1,\dots,s_t)$ a fixed cylinder and put

$$A(m,x) = \sum_{\substack{T^k x \in B(s_1,\dots,s_t) \\ 0 \leq k \leq m-1}} 1$$

Then

$$\lim_{m\to\infty} \frac{A(m,x)}{m} = \mu(B(s_1,\dots,s_t)) \qquad \text{a.e.}$$

Proof: Apply the ergodic theorem to the indicator function of $B(s_1,\dots,s_t)$. We note that $\frac{A(m,x)}{m}$ is the frequence of the block of digits $(s_1,\dots,s_n)$ under the first m+t-1 digits of x.

Next we prove:

Theorem 7.9: There exists a constant $h(T) > 0$ that

$$\lim_{s\to\infty} \frac{1}{s} \log \mu(B(k_1,\dots,k_s)(x)) = -h(T) \qquad \text{a.e.}$$

Note: (i) $B(k_1,\dots,k_s)(x)$ is the cylinder with $x \in B(k_1,\dots,k_s)$

(ii) The constant $h(T)$ can be seen as the entropy of the transformation T (see Billingsley [1] for a treatment of the questions involved; Schweiger [9])

Proof: Put

$$D(x) = \left|\frac{\partial T}{\partial x}(x)\right| , \quad x \in B$$

the absolute value of the Jacobian of $T : B \to B$. An easy calculation gives $D(x) = x_1^{-(n+1)}$. From this we see $-\log D(x)$ integrable and application of the ergodic theorem gives

$$\lim_{s\to\infty} \frac{1}{s} \sum_{k=0}^{s-1} \log D(T^k x) = \int_B \log D(x)\,dM = h(T)$$

The chain rule for Jacobian says

$$\sum_{k=0}^{s-1} \log D(T^k x) = \log \left|\frac{\partial T^s}{\partial x}(x)\right|$$

Using $\Delta(k_1,\dots,k_s)(T^s x)\, \frac{\partial T^s}{\partial x}(x) = 1$ for $x \in B(k_1,\dots,k_s)$ and

$$\log d_1 + \log \mu(B(k_1,\dots,k_s)) \le \log \Delta(k_1,\dots,k_s)(T^s x) \le$$
$$\le \log d_2 + \log \mu(B(k_1,\dots,k_s))$$

using lemma 2.6 and theorem 6.4 resp. 7.7 with constants $d_1, d_2 > 0$ we see

$$\lim_{s\to\infty} \frac{1}{s} \log \mu(B(k_1,\dots,k_s)) = - h(T) \qquad \text{a.e.}$$

Corollary 7.1o: Putting $A_o^{(n+s)}(x) = A_o^{(n+s)}(k_1(x),\dots,k_s(x))$ we obtain

$$\lim_{s\to\infty} \frac{n+1}{s} \log A_o^{(n+s)}(x) = h(T) \qquad \text{a.e.}$$

Proof: This follows from lemma 2.4 and lemma 2.6 using $A_o^{(s+n+1)} \geq A_o^{(s+j)}$ for $1 \leq j \leq n$.

§ 8. Kuzmin's Theorem

Since μ is equivalent to λ, it is in particular absolutely continuous with respect to λ. The Radon-Nikodym theorem shows the existence of an integrable function uniquely determined a.e. with

$$\mu(A) = \int_A \rho(x)\, d\lambda$$

In the case n=1 the function $\rho(x)$ is known. One can easily verify the conjecture of Gauss

$$\rho(x) = \frac{1}{\log 2} \cdot \frac{1}{1+x}$$

by a direct calculation. The shape of the density function is unknown for $n \geq 2$.

Lemma 8.1: If $T^s B(a_1,\dots,a_s) \cap B(b_1,\dots,b_{n-1}) \neq \emptyset$, then $B(b_1,\dots,b_{n-1}) \subseteq T^s B(a_1,\dots,a_s)$.

Proof: If $T^s B(a_1,\dots,a_s) \cap B(b_1,\dots,b_{n-1}) \neq \emptyset$ then the sequence $a_1,\dots,a_s$, $b_1,\dots,b_{n-1}$ is admissible and

$$B(a_1,\dots,a_s\, ,\, b_1,\dots,b_{n-1}) \subseteq B(a_1,\dots,a_s)$$

Applying T^s gives the result.

Lemma 8.2: The density ρ is uniquely characterized by the properties

(i) $\int_B \rho = 1$ and ρ is bounded

(ii) $\rho(x) = \sum_k \rho(V(k)(x))\Delta(k)(x)$,

$x \in B(b_1,\dots,b_{n-1})$ and the sum runs over all digits k with $k,b_1,\dots,b_{n-1}$ admissible.

Proof: First let ρ be the density of the invariant measure μ. (i) is clear. To prove (ii) consider the equation

$\mu(B(b_1,\ldots,b_{n-1}) \cap A) = \mu(T^{-1}(B(b_1,\ldots,b_{n-1}) \cap A))$ for any measurable set A. This gives

$$\int_{B(b_1,\ldots,b_{n-1})\cap A} \rho(x)\,d\lambda = \int_{T^{-1}(B(b_1,\ldots,b_{n-1})\cap A)} \rho(x)\,d\lambda =$$

$$= \sum_k \int_{B(k,b_1,\ldots,b_{n-1}))\cap T^{-1}A} \rho(x)\,d\lambda = \sum_k \int_{B(b_1,\ldots,b_{n-1})\cap A} \rho(V(k)x)\Delta(k)(x)\,d\lambda$$

by the change-of-variables theorem. Since A was arbitrary, equation (ii) results. The same reasoning shows that an integrable functions satisfying (i) and (ii) is a density of an invariant measure μ^* which is absolutely continuous to λ. Applying the ergodic theorem to an indicator function c_A we have

$$\lim_{n\to\infty} \frac{1}{n} \sum_{k=0}^{n-1} c_A(T^k x) = E(c_A \,\|\, J)$$

which is $\mu^*(A) = \mu(A)$ by the ergodicity of T with respect to μ^*.

Lemma 8.3: Let ψ_o be an integrable function and define $\psi_\nu, \nu \geq 1$, by

$$\psi_\nu(x) = \sum_k \psi_{\nu-1}(V(k)x)\Delta(k)(x)$$

for $x \in B(b_1,\ldots,b_{n-1})$, where the sum runs over all admissible sequences k, b_1, b_{n-1}.

Then

$$\psi_\nu(x) = \sum_{k_1,\ldots,k_\nu} \psi_o(V(k_1,\ldots,k_\nu)x)\Delta(k_1,\ldots,k_\nu)(x)$$

for $x \in B(b_1,\ldots,b_{n-1})$ and $k_1,\ldots,k_\nu$ go over all admissible sequences $k_1,\ldots,k_\nu,b_1,\ldots,b_{n-1}$.

Proof: Our proof is by induction. The formula is true for $\nu = 1$.

$$\psi_{\nu+1}(x) = \sum_k \psi_\nu(v(k)x)\Delta(k)(x) = \sum_k \sum_{k_1,\ldots,k_\nu} \psi_o(V(k_1,\ldots,k_\nu)V(k)x) \cdot$$

$$\cdot\, \Delta(k_1,\ldots,k_\nu)(V(k)x)\Delta(k)(x) =$$

$$= \sum_{k_1,\ldots,k_\nu,k} \psi_o(V(k_1,\ldots,k_\nu,k)x)\Delta(k_1,\ldots,k_\nu,k)(x)$$

Remark: If we take $\psi_o = \rho$, then clearly $\psi_\nu = \rho$ for all $\nu \geq 1$.

Lemma 8.4: If ψ_o is bounded, then we have

$$\int_B \psi_\nu(x)d\lambda = \int_B \psi_o(x)d\lambda$$

Proof: $\int_B \psi_\nu(x)d\lambda = \sum_{b_1,\ldots,b_{n-1}} \int_{B(b_1,\ldots,b_{n-1})} \psi_\nu(x)d\lambda =$

$$= \sum_{b_1,\ldots,b_{n-1}} \sum_{k_1,\ldots,k} \int_{B(b_1,\ldots,b_{n-1})} \psi_o(V(k_1,\ldots,k_\nu)x).\Delta(k_1,\ldots,k_\nu)(x)d\lambda =$$

$$= \sum_{b_1,\ldots,b_{n-1}} \sum_{k_1,\ldots,k_\nu} \int_{B(k_1,\ldots,k_\nu,b_1,\ldots,b_{n-1})} \psi_o(x)d\lambda =$$

$$= \int_B \psi_o(x)d\lambda$$

Note that one must be carefully in the order of summation. The integral and the summation sign can be changed since ψ_o and hence ψ_ν (as can be seen from lemma 8.3) are bounded.

Lemma 8.5: $|\Delta(k_1,\ldots,k_\nu((x) - \Delta(k_1,\ldots,k_\nu)(y)| \leq c_o\lambda(B(k_1,\ldots,k_\nu))d(x,y)$ where $c_o > 0$ is an absolute constant and $d(x,y)$ denotes the Euclidean distance in $\mathbb{R}^n$.

Proof: We know

$$\Delta(k_1,\ldots,k_\nu)(x) = \frac{1}{(A_o^{(\nu+n+1} + \sum_{j=1}^{n} A_o^{(\nu+j)} x_j)^{n+1}}$$

Differentiation and the theorem of finite increments gives the result together with lemma 2.6.

Lemma 8.6: Let ψ_o satisfy

(i) $0 < m \leq \psi_o \leq M$

(ii) $|\psi_o(x) - \psi_o(y)| \leq Kd(x,y)$ for $x,y \in B$ with constants $m,M,K>0$.

Then there exist real constants m_1,M_1,K_1 and K_2 with

(i_1) $0 < m_1 \; \psi_\nu \leq M_1$

(ii_1) $|\psi_\nu(x) - \psi_\nu(y)| \leq K_1 d(x,y,) + K_2\sigma(\nu)$ if $x,y \in B(b_1,\ldots,b_{n-1})$.

(Note: The constants m_1,M_1,K_1 and K_2 do not depend on x,y or ν)

Proof: Theorem 6.4 shows $0 < c_1 \leq \rho(x) \leq c_2$ for the density of the invariant measure μ. Hence $0 < c_3\rho \leq \psi_o \leq c_4\rho$ and $0 < c_3\rho \leq \psi_\nu \leq c_4\rho$ applying lemma 8.3 and lemma 8.2.

This proves (i_1). To show (ii_1) we look at

$$\psi_\nu(x) - \psi_\nu(y) = \sum_{k_1,\ldots,k_\nu} \psi_o(V(k_1,\ldots,k_\nu)CX))\Delta(k_1,\ldots,k_\nu)(x) \quad -$$

$$- \sum_{k_1,\ldots,k_\nu} \psi_o(V(k_1,\ldots,k_\nu)(y)\Delta(k_1,\ldots,k_\nu)(y)$$

Since x and y belong to the same cylinder $B(b_1,\ldots,b_{n-1})$ the summation goes over the same set of digits and hence $\psi_\nu(x) - \psi_\nu(y) =$

$$\sum_{k_1,\ldots,k_\nu} \Big[\psi_o(V(k_1,\ldots,k_\nu)x)\Delta(k_1,\ldots,k_\nu)(x) -$$

$$- \psi_o(V(k_1,\ldots,k_\nu)y)\Delta(k_1,\ldots,k_\nu)(y)\Big] =$$

$$= \sum_{k_1,\ldots,k_\nu} \Big[\psi_o(V(k_1,\ldots,k_\nu)x)\left[\Delta(k_1,\ldots,k_\nu)(x) - \Delta(k_1,\ldots,k_\nu)(y)\right] +$$

$$+ \Delta(k_1,\ldots,k_\nu)(y)\ \left[\psi_o(V(k_1,\ldots,k_\nu)(x) - \psi_o(V(k_1,\ldots,k_\nu)(y)\right]\Big]$$

Using lemma 8.5 and

$$\left|\psi_o(V(k_1,\ldots,k_\nu)y) - \psi_o(V(k_1,\ldots,k_\nu)x)\right| \leq$$

$\leq Kd(V(k_1,\ldots,k_\nu)y\ ,\ V(k_1,\ldots,k_\nu)x) \leq K_2\sigma(\nu)$ since $V(k_1,\ldots,k_\nu)y$ and $V(k_1,\ldots,k_\nu)x$ are both contained in $B(k_1,\ldots,k_\nu)$.

Theorem 8.7: Let ψ_o as in lemma 8.6 and the sequence $(\psi_\nu)_{\nu\varepsilon N}$ defined as in lemma 8.3. Then

$$\left|\psi_\nu - a\rho\right| < b\sigma(\nu)$$

where $b = b(m,M,K)$ and $a = \int_B \psi_o$ are constant.

Proof: From lemma 8.6 (i_1) we see that there are constants $0 < g_o \leq G_o$ with

$$g_o\psi_\nu \leq \psi_{\nu+\mu} \leq G_o\psi_\nu$$

uniformly in ν and μ. In the sequel we may assume $\nu \geq n-1$.
We define for μ fixed:

$$\phi_\nu = \psi_{\mu+\nu} - g_o\psi_\nu$$

$$\zeta_\nu = G_o\psi_\nu - \psi_{\mu+\nu}$$

We set $P_\nu = \bigcup_{a_1,\ldots,a_\nu} B(a_1,\ldots,a_\nu)$ the union of proper cylinders of order ν. A cylinder $B(a_1,\ldots,a_\nu)$ is proper if and only if $B(a_{\nu-n+2},\ldots,a_\nu)$ is proper. Hence

$$P_\nu = T^{-(\nu-n+1)}\ P_{n-1}$$

and

$$\mu(P_\nu) = \mu(P_{n-1})$$

Therefore

$$\lambda(P_\nu) \geq K_3 > 0 \qquad \text{for all } \nu \geq n-1.$$

There exists $x_\nu \in B(a_1,\dots,a_\nu)$ such that

$$\int_{B(a_1,\dots,a_\nu)} \phi_o(x)\,d\lambda \leq \psi_o(x_\nu)\lambda(B(a_1,\dots,a_\nu))$$

and therefore

$$\int_{P_\nu} \phi_o(x)\,d\lambda \leq \Sigma\, \phi_o(x_\nu)\lambda(B(a_1,\dots,a_\nu))$$

Similarly there are x_ν' such that

$$\int_{P_\nu} \zeta_o(x)\,d\lambda = \Sigma\, \zeta_o(x_\nu')\lambda(B(a_1,\dots,a_\nu))$$

We further see (note that lemma 8.3 also applies to the sequences $(\phi_\nu)_{\nu\varepsilon N}$ and $(\zeta_\nu)_{\nu\varepsilon N}$ for $x \varepsilon B(b_1,\dots,b_{n-1})$):

$$\phi_\nu(x) = \sum_{k_1,\dots,k_\nu} \phi_o(V(k_1,\dots,k_\nu)x)\Delta(k_1,\dots,k_\nu)(x) \geq$$

$$\geq \sum_{a_1,\dots,a_\nu} \phi_o(V(a_1,\dots,a_\nu)x)\Delta(a_1,\dots,a_\nu)(x)$$

Here the last sum goes over all $a_1,\dots,a_\nu$ for which $B(a_1,\dots,a_\nu)$ is proper, hence $a_1,\dots,a_\nu$, $b_1,\dots,b_{n-1}$ is admissible. Similarly

$$\zeta_\nu(x) \geq \sum_{a_1,\dots,a_\nu} \phi_o(V(a_1,\dots,a_\nu)x)\Delta(a_1,\dots,a_\nu)(x)$$

Using lemma 2.6 we see

$$\phi_\nu(x)-(n+1)^{-n-1}\int_{P_\nu} \phi_o(x)\,d\lambda \geq \sum_{a_1,\dots,a_\nu} \phi_o(V(a_1,\dots,a_\nu)x)\Delta(a_1,\dots,a_\nu)(x) -$$

$$-\ (n+1)^{-n-1} \sum_{a_1,\dots,a_\nu} \phi_o(x_\nu)\lambda(B(a_1,\dots,a_\nu)) \geq$$

$$\geq (n+1)^{-n-1}\Big[\sum_{a_1,\dots,a_\nu} (\phi_o(V(a_1,\dots,a_\nu)x) - \phi_o(x_\nu)) .\ \lambda(B(a_1,\dots,a_\nu))\Big]$$

Now $\phi_o = \psi_\mu - g_o\psi_o$ is defined (in so far as ψ_μ is concerned) by lemma 8.3 piecewise as a sum on each $B(b_1,\dots,b_{n-1})$. Now $\nu \geq n-1$ and $V(a_1,\dots,a_\nu)x$ and x_ν are contained in the same cylinder $B(a_1,\dots,a_\nu)$, hence they belong both to a same cylinder $B(b_1,\dots,b_{n-1})$ (confer lemma 8.1) and lemma 8.6(ii$_1$) applies:

$$|\,\phi_o(V(a_1,\dots,a_\nu)x) - \phi_o(x_\nu)\,| \leq$$

$$\leq |\,\psi_\mu(V(a_1,\dots,a_\nu)x) - \psi_\mu(x_\nu)\,| + g_o|\psi_o(V(a_1,\dots,a_\nu)x) - \psi_o(x_\nu)\,| \leq$$

$\leq K_1\sigma(\nu) + K_2\sigma(\nu) + g_oK\sigma(\nu)$

since $V(a_1,\dots,a_\nu)x$ and x_ν are both in $B(a_1,\dots,a_\nu)$.

This shows:

$$\phi_\nu - (n+1)^{-n-1} \int_{P_\nu} \phi_o \geq - K_3(K_1 + K_2 + g_oK)\sigma(\nu)$$

Completely similar is the estimation:

$$\zeta_\nu - (n+1)^{-n+1} \int_{P_\nu} \zeta_o \geq - K_3(K_1 + K_2 + G_oK)\sigma(\nu)$$

We put $\phi_\nu = \psi_{\mu+\nu} - g_o\psi_\nu$ and find:

$$\psi_{\mu+\nu} - g_o\psi_\nu - (n+1)^{-n-1} \int_{P_\nu} (\psi_\mu - g_o\psi_o) \geq - K_3(K_1 + K_2 + g_oK)\sigma(\nu) \quad \text{and}$$

$$\psi_{\mu+\nu} \geq \psi_\nu g_o + (n+1)^{-(n+1)} \int_{P_\nu} (\psi_\mu - g_o\psi_o) - K_3(K_1 + K_2 + g_oK)\sigma(\nu) \geq$$

$$\geq \psi_\nu\left[g_o + \frac{1}{M_1(n+1)^{n+1}} \int_{P_\nu} (\psi_\mu - g_o\psi_o) - \frac{K_3}{m_1}(K_1 + K_2 + g_oK)\sigma(\nu)\right]$$

Hence

$$\psi_{\mu+\nu} \geq g_1\psi_\nu$$

with

$$g_1 = g_o\eta(\nu) + \beta(\nu)$$

$$\eta(\nu) = 1 - \frac{1}{M_1(n+1)^{n+1}} \int_{P_\nu} \psi_o - \frac{K_3K}{m_1}\sigma(\nu)$$

$$\beta(\nu) = \frac{1}{M_1(n+1)^{n+1}} \int_{P_\nu} \psi_\mu - \frac{K_3}{m_1}(K_1 + K_2)\sigma(\nu)$$

In a completely analoguous fashion we obtain

$$\psi_{\mu+\nu} \leq G_1\psi_\nu$$

with

$$G_1 = G_o\theta(\nu) + \delta(\nu)$$

$$\theta(\nu) = 1 - \frac{1}{M_1(n+1)^{n+1}} \int_{P_\nu} \psi_o + \frac{K_3K}{m_1}\sigma(\nu)$$

$$\delta(\nu) = \frac{1}{M_1(n+1)^{n+1}} \int_{P_\nu} \psi_\mu + \frac{K_3}{m_1}(K_1 + K_2)\sigma(\nu)$$

We see: The relation $g_o\psi_\nu \leq \psi_{\nu+\mu} \leq G_o\psi_\nu$ implies $g_1\psi_\nu \leq \psi_{\nu+\mu} \leq G_1\psi_\nu$ with g_1, G_1 as described above.

Iteration gives the recursion

$$g_{r+1} = g_r\eta(\nu) + \beta(\nu) \qquad\qquad G_{r+1} = G_r\theta(\nu) + \delta(\nu)$$

If $\nu \geq \nu_o$ the iteration gives the limits

$$g(\nu) = \lim_{r\to\infty} g_r = \frac{\beta(\nu)}{1-\eta(\nu)} \geq Q(\nu,\mu) - K_4\sigma(\nu)$$

$$G(\nu) = \lim_{r\to\infty} G_r = \frac{\delta(\nu)}{1-\Theta(\nu)} \leq Q(\nu,\mu) + K_5\sigma(\nu)$$

where

$$Q(\nu,\mu) = \frac{\int_{P_\nu}\psi_\mu}{\int_{P_\nu}\psi_o}$$

To obtain the last inequalities uniformly in $\nu \geq \nu_o$ we used $\sigma(\nu) \leq \sigma(\nu_o)$ and a uniform estimation for $\int_{P_\nu}\psi_\mu$ and $\int_{P_\nu}\psi_o$ which can be obtained by $0 < m_1 \leq \psi_t \leq M_1$ for all $t \geq 0$ and $K_3 \leq \lambda(P_\nu) \leq 1$.

We now have

$$\left|\psi_{\nu+\mu} - Q(\nu,\mu)\psi_\nu\right| \leq K_6\sigma(\nu)$$

Now we integrate over B and use lemma 8.4 to obtain

$$\left|1 - Q(\nu,\mu)\right| \leq K_6\sigma(\nu)$$

which gives now

$$\left|\psi_{\nu+\mu} - \psi_\nu\right| \leq b_o\sigma(\nu) \quad \text{with a constant } b_o = b_o(m,M,K).$$

This shows that the sequence $(\psi_\nu)_{\nu\in N}$ is fundamental in the space of integrable functions on K. Putting

$$\lim_{\nu\to\infty} \psi_\nu = a\rho^*$$

with $a = \int_B \psi_o(x)d\lambda$ we obtain $(\mu \to \infty)$:

$$\left|a\rho^* - \psi_\nu\right| < b\sigma(\nu) \quad \text{with } b = b(m,M,K).$$ From

$$\psi_{\nu+1}(x) = \sum_k \psi_\nu(V(k)x)\Delta(k)(x), \quad x \in B(b_1,\dots,b_{n-1})$$

and $\quad |\sum_k [\psi_\nu(V(k)x) - \rho^*(V(k)x)]\,\Delta(k)(x)| \le K_7\sigma(\nu) \sum_k \lambda(B(k) \le K_7\sigma(\nu)$

we get

$$\rho^*(x) = \sum_k \rho^*(V(k)x)\Delta(k)(x)$$

By lemma 8.2 ρ^* is the density of μ.

Remark 1: One could replace lemma 8.1 by the following statement:There is a countable partition of B into sets $A_1,\ldots,A_m$ with the property:If $T^sB(a_1,\ldots,a_s) \cap A_j \neq \emptyset$, then $A_j \subseteq T^sB(a_1,\ldots,a_s)$. In fact the A_j are unions of cylinders $B(b_1,\ldots,b_{n-1})$. Using this result (which follows from geometrical considerations on the images $T^sB(a_1,\ldots,a_s)$) one sees that ρ is continuous on each A_j, $1 \le j \le m$.

Remark 2: From theorem 8.7 one can at least in principle approximate ρ. To estimate the error an estimation of $\sigma(\nu)$ would be of great value . For a computer experiment see W.A.Beyer and M.S.Waterman [1].

From theorem 8.7 we now deduce the important

Theorem 8.8: For all $A \in \mathcal{F}$ we have

$$|\lambda_s(A) - \mu(A)| < b\lambda(A)\sigma(s)$$

In particular this implies

$$\lim_{s\to\infty} \lambda_s(A) = \mu(A)$$

Proof: From lemma 6.5 we know that λ_n is absolutely continuous with respect to λ. Hence

$$\lambda_m(A) = \lambda(T^{-m}A) = \int_A \psi_m(x)\,d\lambda$$

One calculates

$$\lambda_{m+1}(A) = \lambda_m(T^{-1}A) = \int_{T^{-1}A} \psi_m(x)\,d\lambda =$$

$$= \sum_{b_1,\ldots,b_{n-1}} \int_{T^{-1}(A\cap B(b_1,\ldots,b_{n-1}))} \psi_m(x)\,d\lambda =$$

$$= \sum_{b_1,\ldots,b_{n-1}} \sum_k \int_{A\cap B(b_1,\ldots,b_{n-1})} \psi_m(V(k)x)\Delta(k)(x)\,d\lambda =$$

$$= \sum_{b_1,\ldots,b_{n-1}} \int_{A\cap B(b_1,\ldots,b_{n-1})} \sum_k \psi_m(V(k)x)\Delta(k)(x)\,d\lambda$$

Hence $\psi_{m+1}(x) = \sum_k \psi_m(V(k)x)\Delta(k)(x)$ for $x \in B(b_1,\ldots,b_{n-1})$ and since clearly $\psi_0(x) \equiv 1$ satisfies the assumptions of theorem 8.7 we deduce

$$|\psi_s - \rho| < b\sigma(s)$$

Integration yields the result.

Note 1: It should be pointed out that the proof of Kuzmin's theorem for continued fractions as presented in the book of Khintchine [1] not only gives a weaker result but contains a gap: The value of n to insure monotonicity of the sequences g_r and G_r possibly increases with increasing r. This gap can be found in a series of subsequent papers (Schweiger [11], Tran-Vinh-Hien [1], Waterman [1],[2]). The proof presented here is a variant of Schweiger-Waterman [1]. The easier case of continued fractions is also handled in P.Levy [1], P.Szüsz [1],[2] and F.Schweiger [7].

Note 2: In § 9 we will show that $\sigma(\nu)$ is exponentially decreasing. In [1] Gordin announces a generalized Kurzmin's theorem with an exponentially fast convergence. The methods seem to use functional analysis.

§ 9. Convergence results

In lemma 2.9 we have shown: $\sigma(m) \to 0$ with $m \to \infty$.
In view of the theorems presented in § 8 it would be of great value to estimate the speed of convergence.

Theorem 9.1: For $n = 1$ we have $\sigma(\nu) = O(\theta^{-2\nu})$ where θ is the unique root of $F(x) = x^2 - x - 1 = 0$ with $1 < \theta$. This result is best possible.

Proof: Clearly for $n = 1$ the Jacobi algorithm reduces to the continued fraction algorithm. By lemma 1.2 we have

$$\alpha_1 = \frac{A_1^{(s+2)} + y_1 A_1^{(s+1)}}{A_o^{(s+2)} + y_1 A_o^{(s+1)}}, \qquad y = T^s\alpha$$

and therefore

$$\left|\alpha_1 - \frac{A_1^{(s+2)}}{A_o^{(s+2)}}\right| \le \frac{|y_1(A_1^{(s+1)}A_o^{(s+2)} - A_o^{(s+1)}A_1^{(s+2)}|}{(A_o^{(s+2)} + y_1 A_o^{(s+1)})\, A_o^{(s+2)}} \le \frac{1}{(A_o^{(s+2)})^2}$$

Here we used $0 \le y_1 < 1$ and $|A_1^{(s+1)}A_o^{(s+2)} - A_o^{(s+1)}A_1^{(s+2)}| = 1$.

Since $A_o^{(2)} = 1$, we have

$$A_o^{(s+2)} = A_o^{(s)} + k_1^{(s)} A_o^{(s+1)} \ge (\theta^s + \theta^{s+1}) = \theta^{s+2} \quad \text{by induction.}$$

Hence

$$A_o^{(s+2)} \ge \theta^{s+2} \qquad \text{and} \qquad (A_o^{(s+2)})^{-1} \le \theta^{-s-2}$$

therefore

$$\alpha_1 - \frac{A_1^{(s+2)}}{A_o^{(s+2)}} = O(\theta^{-2s})$$

Since $\alpha_1 \in B(k_1,\dots,k_{s-1})$ was arbitrary, we see $\sigma(s) = O(\theta^{-2s})$.

For $\alpha_1 = \psi\,[1,1,1,\dots]$, we have $A_o^{(s+2)} = A_o^{(s)} + A_o^{(s+1)}$ and therefore

$A_o^{(s+2)} = O(\theta^{s+2})$.

This shows the result best possible (in the order of approximation) .

Theorem 9.2 (Paley-Ursell [1]):

For $n = 2$ we have $\sigma(\nu) = O(\theta^{-\nu})$ where θ is the unique root

$F(x) = x^3 - x^2 - 1 = 0$ with $1 < \theta$.

Proof:

We again use

$$\alpha_i = \frac{A_i^{(s+3)} + y_1 A_i^{(s+1)} + y_2 A_i^{(s+2)}}{A_o^{(s+3)} + y_1 A_o^{(s+1)} + y_2 A_o^{(s+2)}}$$

$$i = 1,\ 2 \qquad ; \qquad y = T^s\alpha$$

We will show

$$\left| \alpha_i - \frac{A_i^{(s+3)}}{A_o^{(s+3)}} \right| = O(\theta^{-s}) \quad , \quad i = 1,\ 2$$

Then clearly $\sigma(s) = O(\theta^{-s})$.

For sake of explicitness we take $i = 1$.

We define

$$[s+1,\ s+2] = \begin{vmatrix} A_1^{(s+1)} & A_1^{(s+2)} \\ A_o^{(s+1)} & A_o^{(s+2)} \end{vmatrix}$$

$$[s\ ,\ s+2] = \begin{vmatrix} A_1^{(s)} & A_1^{(s+2)} \\ A_o^{(s)} & A_o^{(s+2)} \end{vmatrix}$$

This gives the recursion formulas

$$[s+2,\ s+3] = -\,|s,\ s+2| - a_1^{(s)}\,[s+1,\ s+2]$$

$$[s+1,\ 1+3] = -\,|s,\ s+1| + a_2^{(s)}\,[s+1,\ s+2] \ .$$

Here we used

$$A_j^{(s+3)} = A_j^{(s)} + a_1^{(s)} A_j^{(s+1)} + a_2^{(s)} A_j^{(s+2)} , \quad 0 \leq j \leq 2$$

One calculates easily

$$\left|\alpha_1 - \frac{A_1^{(s+3)}}{A_0^{(s+3)}}\right| \leq \frac{|[s+1,s+3]| + |[s+2,s+3]|}{(A_0^{(s+3)})^2}$$

Our theorem can be derived easily if we can show

$$|[s+1, s+3]| \leq A_0^{(s+3)}$$

$$|[s+2, s+3]| \leq A_0^{(s+3)}$$

We postpone the proof of these crucial relations and will deduce first the theorem: Since $A_0^{(3)} = 1$ we can prove using

$$A_0^{(s+3)} = A_0^{(s)} + a_1^{(s)} A_0^{(s+1)} + a_2^{(s)} A_0^{(s+2)} \geq A_0^{(s)} + A_0^{(s+2)} \geq \theta^s + \theta^{s+2} = \theta^{s+3}$$

the result by induction. It is known that the estimate

$$\alpha_i - \frac{A_i^{(s+3)}}{A_0^{(s+3)}} = O\left(\frac{1}{A_0^{(s+3)}}\right)$$

is best possible (see Paley-Ursell [1]), but their construction does not imply $\sigma(s) = O(\theta^{-s})$ best possible.

To prove the basic inequalities we will proceed by induction. A lengthy calculation shows the inequalities true for $0 \leq s \leq 2$. We now assume the inequalities true for all t with $0 \leq t \leq s-1$ and we will deduce the assertion for s.

<u>Lemma 9.3</u>: If $[s, s+2] \quad [s+1,s+2] \geq 0$, then

$$\min(|[s, s+2]|, |[s+1,s+2]|) \leq A_0^{(s+1)}$$

<u>Proof</u>:

(1) $[s,s+2] \geq 0$, $[s+1,s+2] \geq 0$

If $[s,s+1] \geq 0$, we have

$$0 \leq [s+1,s+2] = -[s-1,s+1] - a_1^{(s-1)} [s,s+1] \leq -[s-1,s+1] \leq A_0^{(s+1)}$$

If $[s,s+1] \leq 0$ we have

$$0 \leq [s,s+2] = -[s-1,s] + a_2^{(s-1)} [s,s+1] \leq -[s-1,s] \leq A_0^{(s)} \leq A_0^{(s+1)}$$

(2) $[s,s+2] \leq 0$, $[s+1,s+2] \leq 0$

If $[s,s+1] \geq 0$ we get

$$0 \geq [s,s+2] = -[s-1,s] + a_2^{(s-1)}[s,s+1] \geq [-s-1,s] \geq -A_0^{(s)} \geq -A_0^{(s+1)}$$

If $[s,s+1] \leq 0$ we see

$$0 \geq [s+1,s+2] \geq -[s-1,s+1] - a_1^{(s-1)}[s,s+1] \geq -[s-1,s+1] \geq -A_0^{(s+1)}$$

Lemma 9.4: $|[s+2,s+3]| \leq A_0^{(s+3)}$

Proof:

(1) We first assume $[s,s+2]\ [s+1,s+2] \leq 0$

From $[s+2,s+3] = -[s,s+2] - a_1^{(s)}[s+1,s+2]$ we get

$$|[s+2,s+3]| \leq \max(|[s,s+2]|,\ a_1^{(s)}|[s+1,s+2]|) \leq$$

$$\leq \max(A_0^{(s+2)},\ a_1^{(s)}A_0^{(s+2)}) \leq a_2^{(s)}A_0^{(s+2)} \leq A_0^{(s+3)}$$

Here we used $a_1^{(s)} \leq a_2^{(s)}$

(2) $[s,s+2]\ [s+1,s+2] \geq 0$

From the recursion formula we obtain

$$|[s+2,s+3]| \leq |[s,s+2]| + a_1^{(s)}|[s+1,s+2]|$$

Using lemma 9.3 we have two possibilities:

(2.1) $|[s,s+2]| \leq A_0^{(s+1)}$

Clearly

$|[s+1,s+2]| \leq A_0^{(s+2)}$ by induction hypothesis and

$$|[s+2,s+3]| \leq A_0^{(s+1)} + a_1^{(s)}A_0^{(s+2)}$$

If $a_1^{(s)} = 0$ we obtain $|[s+2,s+3]| \leq A_0^{(s+1)} \leq A_0^{(s+3)}$

If $a_1^{(s)} \geq 1$ we conclude as follows:

$$|[s+2,s+3]| \leq A_0^{(s+1)} + a_1^{(s)}A_0^{(s+2)} \leq a_1^{(s)}A_0^{(s+1)} + a_2^{(s)}A_0^{(s+2)} \leq A_0^{(s+3)}$$

(2.2) $|[s+1,s+2]| \leq A_0^{(s+1)}$

Then by induction

$$|[s+2,s+3]| \leq A_0^{(s+2)} + a_1^{(s)}A_0^{(s+1)} \leq a_2^{(s)}A_0^{(s+2)} + a_1^{(s)}A_0^{(s+1)} \leq A_0^{(s+3)}$$

Lemma 9.5: $|[s+1,s+3]| \leq A_0^{(s+3)}$

Proof: We start with the recursion formula

$$[s+1,s+3] = -[s,s+1] + a_2^{(s)}[s+1,s+2]$$

Let us first assume $a_1^{(s)} \geq 1$. Then we have

$$|[s+1,s+3]| \leq |[s,s+1]| + a_2^{(s)}|[s+1,s+2]| \leq A_0^{(s+1)} + a_2^{(s)}A_0^{(s+2)} \leq$$

$$\leq a_1^{(s)}A_0^{(s+1)} + a_2^{(s)}A_0^{(s+2)} \leq A_0^{(s+3)}$$

The worst case is $a_1^{(s)} = 0$. Then the conditions of the algorithm imply:

$$a_1^{(s-1)} < a_2^{(s-1)}$$

We calculate

$[s+1,s+3] = -[s,s+1] + a_2^{(s)} [s+1,s+2] = - [s,s+1] + a_2^{(s)} (- [s-1,s+1] -$

$- a_1^{(s-1)}[s,s+1]) = [s,s+1] (-1- a_1^{(s-1)} a_2^{(s)} - a_2^{(s)} - a_2^{(s)} [s-1,s+1]$

(1) $\quad [s,s+1] [s-1,s+1] \leq 0$

Then

$|[s+1,s+3]| \leq \max (|[s,s+1]| (1 + a_1^{(s-1)} a_2^{(s)}), a_2^{(s)} |[s-1,s+1]|)$

Now

$$a_2^{(s)} |[s-1,s+1]| \leq a_2^{(s)} A_o^{(s+1)} \leq a_2^{(s)} A_o^{(s+2)} \leq A_o^{(s+3)}$$

Since $\quad 1 + a_1^{(s-1)} \leq a_2^{(s-1)}$ in our case clearly

$$1 + a_1^{(s-1)} a_2^{(s)} \leq (1 + a_1^{(s-1)}) a_2^{(s)} \leq a_2^{(s-1)} a_2^{(s)}$$

and therefore

$|[s,s+1]| (1 + a_1^{(s-1)} a_2^{(s)}) \leq A_o^{(s+1)} a_2^{(s-1)} a_2^{(s)} \leq A_o^{(s+3)}$

(here we used $A_o^{(s+3)} \geq a_2^{(s)} A_o^{(s+2)} \geq a_2^{(s-1)} a_2^{(s)} A_o^{(s+1)}$)

(2) $\quad [s,s+1] [s-1,s+1] \geq 0$

From lemma 9.3 we obtain $\min (|[s,s+1]|, |[s-1,s+1]|) = A_o^{(s)}$

If $|[s,s+1]| \leq A_o^{(s)}$, we see (using induction):

$|[s+1,s+3]| \leq |[s,s+1]| (1 + a_1^{(s-1)} a_2^{(s)}) + |[s-1,s+1]| a_2^{(s)} \leq$

$\leq A_o^{(s)} (1 + a_1^{(s-1)} a_2^{(s)}) + a_2^{(s)} A_o^{(s+1)} \leq A_o^{(s+3)}$

The last inequality is from

$A_o^{(s+3)} = A_o^{(s)} + a_2^{(s)} A_o^{(s-1)} + a_2^{(s)} a_1^{(s-1)} A_o^{(s)} + a_2^{(s)} a_2^{(s-1)} A_o^{(s+1)}$

If $|[s-1,s+1]| \leq A_o^{(s)}$ we must again discuss two cases:

(2.1) $\quad a_1^{(s-1)} \geq 1$

Then we get

$|[s+1,s+3]| \leq A_o^{(s+1)} (1 + a_1^{(s-1)} a_2^{(s)}) + a_2^{(s)} A_o^{(s)} \leq A_o^{(s+1)} a_2^{(s-1)} a_2^{(s)} +$

$+ a_1^{(s-1)} a_2^{(s)} A_o^{(s)} \leq A_o^{(s+3)}$

(2.2) $\quad a_1^{(s-1)} = 0$

Then again $\; a_1^{(s-2)} < a_2^{(s-2)}$.

$[s+1,s+3] = -[s,s+1] - a_2^{(s)}[s-1,s+1] = [s-2,s] + a_1^{(s-2)}[s-1,s] - a_2^{(s)}[s-1,s+$

$$\left|[s+1,s+3]\right| \le A_o^{(s)}(1 + a_1^{(s-2)}) + A_o^{(s)} a_2^{(s)} \le A_o^{(s)} a_2^{(s-2)} + A_o^{(s)} a_2^{(s)} \le$$

$$\le A_o^{(s)} + A_o^{(s)} a_2^{(s)} a_2^{(s-2)} \le A_o^{(s+3)}$$

We used $a_2^{(s-2)} \ge 1$ and $a_2^{(s)} \ge 1$.

In the general case, we prove

Theorem 9.6 (Fischer [1]): For $n \ge 1$ we have $\sigma(\nu) = O(\theta^{-\nu})$, where $\theta^n = 1 - (n+1)^{-n}$.

Proof: We start with the relation

$$\frac{A_i^{(s+n+1)}}{A_o^{(s+n+1)}} = \sum_{j=o}^{n} \lambda_j \frac{A_i^{(s+j)}}{A_o^{(s+j)}}$$

where $\lambda_j = k_j^{(n)} A_o^{(s+j)} / A_o^{(s+n+1)}$ (as usual we put $k_o^{(n)} = 1$) .

Clearly $\lambda_n \ge (n+1)^{-1}$ and $\sum_{j=1}^{n} \lambda_j = 1$. With the help of an easy induction we obtain

$$\frac{A_i^{(s+n+g)}}{A_o^{(s+n+g)}} = \sum_{j=o}^{n} \lambda_j^{(g)} \frac{A_i^{(s+j)}}{A_o^{(s+j)}}$$

for any $g \ge 1$ and the coefficients satisfy $\lambda_j^{(g)} \ge 0$, $\lambda_n^{(g)} \ge (n+1)^{-g}$, $\sum_{j=o}^{n} \lambda_j^{(g)} = 1$. We now estimate

$$\left| \frac{A_i^{(s+n+g)}}{A_o^{(s+n+g)}} - \frac{A_i^{(s+n)}}{A_o^{(s+n)}} \right| = \left| \sum_{j=o}^{n-1} \lambda_j^{(g)} \left(\frac{A_i^{(s+j)}}{A_o^{(s+j)}} - \frac{A_i^{(s+n)}}{A_o^{(s+n)}} \right) \right| \le (1-(n+1)^{-n})\sigma(s).$$

For any pair g,h with $1 \le g < h \le n$, we have (using $h = g + (h - g)$) :

$$\left| \frac{A_i^{(s+n+g)}}{A_o^{(s+n+g)}} - \frac{A_i^{(s+n+h)}}{A_o^{(s+n+h)}} \right| \le (1 - (n+1)^{-n}) \sigma(s+g) \le$$

$$\le (1 - (n+1)^{-n}) \sigma(s)$$

Therefore

$$\sigma(s+n) \le (1 - (n+1)^{-n}) \sigma(s) .$$

Since $\alpha_i = \sum_{j=o}^{n} f_j^{(s)} A_i^{(s+j)} / A_o^{(s+j)}$ with $0 \leq f_j^{(s)} \leq 1$ and $\sum_{j=o}^{n} f_j^{(s)} = 1$ (see lemma 2.9), this result gives a quick convergence proof for Jacobi algorithm.

Corollary 9.7: $\sum_{s=1}^{\infty} \sigma(s) < \infty$.

This result will be important in the next chapter.

There is a lot of convergence results for periodic algorithms (Perron [1] Bernstein [1]). An interesting relation to the "Littlewood conjecture" was given by Bottorf [1]. The "Littlewood conjecture" claims: Given two irrational numbers u_1, u_2, there exists an infinite sequence of triples of integers p,q,r such that $|q||u_1 q - p||u_2 q - r|$ is arbitrarily small.

We prove

Theorem 9.10 (Bottorf): Let $1 \leq a^2 < 4b$, a,b integers and $\sigma > 1$ the unique real root of $F(x) = x^3 - bx^2 - ax - 1 = 0$. The pair $(\sigma^{-1}, \sigma - b)$ satisfies the Littlewood conjecture.

Proof: The condition $1 \leq a^2 < 4b$ implies $a \leq b$. Hence $b < \sigma < b + 1$ and the point $(\sigma^{-1}, \sigma - b)$ has the purely periodic algorithm $k_1^{(s)} = a$, $k_2^{(s)} = b$, $S \geq 1$.

Since the discriminant

$$\Delta = a^2b^2 + 4a^3 - 4b^3 - 18ab - 27 < 0,$$

the two other roots α, β are complex and satisfy $\bar{\alpha} = \beta$.

The difference equation

$$y(n+3) - by(n+2) - ay(n+1) - y(n) = 0$$

has the general solution

$$y(n+1) = A\sigma^n + B\alpha^n + C\beta^n$$

The differences

$$H_1(n+1) = A_o^{(n+1)} \sigma^{-1} - A_1^{(n+1)}$$

$$H_2(n+1) = A_o^{(n+1)} (\sigma - b) - A_2^{(n+1)}$$

are solutions. Using the initial values $H_1(1) = -1$, $H_1(2) = 0$, $H_1(3) = \sigma^{-1}$ and $H_2(1) = 0$, $H_2(2) = -1$, $H_2(3) = \sigma - b$ we calculate

$$H_1(n+1) = -B_1\alpha^n + \bar{B}_1\bar{\alpha}^n$$

$$H_2(n+1) = -B_2\alpha^n + \bar{B}_2\bar{\alpha}^n$$

We put $B_1 = |B_1| e^{i\theta}$, $B_2 = |B_2| e^{i\eta}$ and $\alpha = |\alpha| e^{i\psi}$ and obtain

$$H_1(n+1) = 2|B_1| \sigma^{-\frac{n}{2}} \sin(\theta+n\psi)$$

$$H_2(n+1) = 2|B_2| \sigma^{-\frac{n}{2}} \sin(\eta+n\psi)$$

Therefore using $|\sin(\eta+n\psi)| \leq 1$ we obtain $|H_1(n+1)\, H_2(n+1)| \leq$ $\leq 4|B_1||B_2| \sigma^{-n} |\sin(\theta+n\psi)|$ Since $A_o^{(n+1)}$ is a solution of the difference equation, we have

$$\lim_{n\to\infty} \frac{A_o^{(n+1)}}{\sigma^n} = \frac{\beta - \alpha}{i\sqrt{\Delta}}$$

This shows

$$|H_1(n+1)|\, |H_2(n+1)| = |\sin(\theta+n\psi)|\, O\left(\frac{1}{A_o^{(n+1)}}\right)$$

Given θ and ψ the inequality

$$\left|\frac{\psi}{\pi} n - m + \frac{\theta}{\pi}\right| < \frac{3}{n}$$

has infinitely many integral solutions (n,m) with $n > 0$ (see e.g. Koksma [1], p.76)

Therefore

$$|\sin(\theta+n\psi)| = |\sin(\theta+n\psi - m\pi)| < \varepsilon$$

infinitely often.

§ 1o. The Borel-Cantelli lemma of Schmidt-Philipp

In [2] Philipp proved a very useful theorem of which we give a slightly extended version. Another more general form was given by Ennola [1] .

Theorem 1o.1: Let $(\Omega,\mathcal{F},P)$ be a probability space and $(E_n)_{n\in N}$ a sequence of measurable sets $E_n \in \mathcal{F}$. For each $x\in\Omega$ define $A(N,x) = \sum_{n\leq N} C_{E_n}(x)$ and define further $\phi(N) = \sum_{n\leq N} P(E_n)$. Suppose there exists a convergent series $\Sigma\alpha_k$ with $\alpha_k \geq 0$ such that

$$P(E_n \cap E_{n+m}) \leq P(E_n)P(E_{n+m}) + \left[P(E_{n+m}) + P(E_n)\right]\alpha_m + P(E_{n+m})\alpha_n$$

Then $A(N,x) = \phi(N) + O(\phi^{\frac{1}{2}}(N) \log^{\frac{3}{2}+\varepsilon} \phi(N))$ for every $\varepsilon > 0$ for almost all x.

<u>Proof</u>: We follow the line of the before mentioned paper of Philipp whic traces back to W.M.Schmidt [2]. We first assume $\sum_{n=1}^{\infty} P(E_n)$ converges. Then for every $\eta > 0$ we can find an index $n(\eta)$ with $\sum_{n \geq n(\eta)} P(E_n) < \eta$. Hence $P(\bigcup_{n \geq n(\eta)} E_n) < \eta$ and $P(\bigcap_{k=1}^{\infty} \bigcup_{n \geq k} E_n) = 0$. Therefore $A(N,x)$ is finite almost everywhere and the theorem is clear. So we assume $\sum_{n=1}^{\infty} P(E_n)$ diverges. We put $A(n,n+m,x) = \sum_{i=n+1}^{n+m} c_{E_i}(x) = A(n+m,x) - A(n,x)$ and

$$\phi(n,n+m) = \int_{\Omega} A(n,n+m,x)\, dP(x)$$

Then clearly $A(0,N,x) = A(N,x)$ and $\phi(0,N) = \phi(N)$

Calculations gives

$$\int_{\Omega} (A,n+m,x) - \phi(n,n+m))^2 dP(x) = \phi(n,n+m) - \phi^2(n,n+m) + 2 \sum_{n+1 \leq i < j \leq n+m} P(E_i \cap E_j) \leq$$

$$\leq \phi(n,n+m) - \phi^2(n,n+m) + 2 \sum_{i<j} P(E_i)P(E_j) + 2 \Big[\sum_{i<j} P(E_i)\alpha_{j-i} + \sum_{i<j} P(E_j)\, \alpha_i +$$

$$+ \sum_{i<j} P(E_j)\, \alpha_{j-i} \Big]$$

Now $\phi(n,n+m)^2 = (\sum_{n+1 \leq i \leq n+m} P(E_i))^2$ and therefore

$$- \phi(n,n+m)^2 + 2 \sum_{i<j} P(E_i)\, P(E_j) = - \sum_{i=n+1}^{n+m} P(E_i)^2 \leq 0$$

Furthermore $\sum_{i<j} P(E_i)\alpha_{j-i} \leq \sum_{n+1 \leq i \leq n+m} P(E_i) \sum_{i<j \leq n+m} \alpha_{j-i} = O(\sum_{n+1 \leq i \leq n+m} P(E_i))$

$$\sum_{i<j} P(E_j)(\alpha_i + \alpha_{j-i}) \leq \sum_{j} P(E_j) \sum_{i<j} (\alpha_i + \alpha_{j-i}) = O(\sum_{n+1 \leq i \leq n+m} P(E_i))$$

Both estimates follow from $\sum_{n=1}^{\infty} \alpha_n \leq A$ with suitable A.

Hence we obtain $\int_{\Omega} (A(n,n+m,x) - \phi(n,n+m))^2 dP(x) = O(\phi(n,n+m))$

For integer $u \geq 0$ we define N_u to be the largest integer N with $\phi(N) < u$

Denote by L_r the set of intervalls $]u,v] =]t.2^s, (t+1)2^s]$ where $s \geq 0$, $t \geq 0$ integer and $v \leq 2^r$ with $r \geq 1$. For fixed $s \leq r$ the intervalls $]t.2^s, (t+1)2^s]$ from a disjoint covering of $]0,2^r]$ if t runs from 0 to $2^{r-s} - 1$.

Since $\phi(N_u,N_v) = \sum_{N_u < i \leq N_v} P(E_i)$ we obtain

$$\Sigma\ \phi(N_u,N_v) \leq \phi(N_{2^r}) < 2^r$$

where the summation is extended over all $]u,v]$ which cover $]0,\ 2^r]$ for s fixed. From $s \leq r$ we find

$$\sum_{]u,v] \in L_r} \phi\ (N_u,N_v) < (r+1)2^r$$

Put
$$Z(r,x) = \sum_{]u,v] \in L_r} (A(N_u,N_v,x) - \phi(N_u,N_v))^2$$

The estimates just proved yield

$$\int_\Omega Z(r,x)\ dP(x) = O(\Sigma\ \phi(N_u,N_v)) = O(r\ 2^r)$$

This shows $\int_\Omega \frac{Z(r,x)}{2^r r^{2+\varepsilon}}\ d\ P(x) = O\ (r^{-1-\varepsilon})$

Hence
$$\sum_{r=1}^{\infty} \int_\Omega \frac{Z(r,x)}{2^r r^{2+\varepsilon}}\ d\ P(x) = O\ (1)$$

The theorem of H.Lebesgue-B.Levi (see e.g. Munroe [1] p.186) shows $\int_\Omega \sum_{r=1}^{\infty} \frac{Z(r,x)}{2^r r^{2+\varepsilon}} = O(1)$. Therefore $Z(r,x) = O(2^r r^{2+\varepsilon})$ almost everywhere.

If w is an integer and $2^{r-1} < w \leq 2^r$ then $]0,w]$ can be represented as an union of at most r+1 intervalls of L_r and the same holds for $]0,N_w]$. Hence $A(N_w,x) - \phi(N_w) = \Sigma(A\ (N_u N_v,x) - \phi(N_u N_v))$ where the sum is over at most r+1 intervalls $]u,v] \in L_r$. With help of the Cauchy-Schwarz inequality we obtain

$$(A\ (N_w,x) - \phi(N_w))^2 \leq O(2^r r^{2+\varepsilon})(r+1) = O(2^r r^{3+\varepsilon})$$

From this we see $A(N_w,x) - \phi(N_w) = O(2^{\frac{r}{2} + \frac{\varepsilon}{2}})$

Since $\phi(N_w) < w \leq \phi(N_w + 1)$ and $2^{r-1} < w \leq 2^r$ we obtain

$$A(N_w,x) - \phi(N_w) = O(\phi^{\frac{1}{2}}\ (N_w)\ \log^{\frac{3}{2} + \frac{\varepsilon}{2}} \phi(N_w))$$

and the theorem is proved for $N = N_w$.

If N is arbitrary we find w with $N_w \leq N < N_{w+1}$ and get

$$A(N_w,x) \leq A(N,x) \leq A(N_{w+1},x)$$

Clearly $\phi(N_w) \leq \phi(N) \leq \phi(N_{w+1})$.

From the definition of N_w follows $\phi(N_w) < w \leq \phi(N_w+1)$

$$\phi(N_{w+1}) < w+1 \leq \phi(N_{w+1} + 1)$$

Therefore

$\phi(N_{w+1}) < \phi(N_w+1) + 1 \leq \phi(N_w) + 2$ (since $P(E_j) \leq 1$ for all E_j). The result is now obvious.

To deduce significant results for Jacobi algorithm we need

Theorem 1o.2: Let $E = B(k_1,\ldots,k_s)$ and $F \in \mathcal{L}$. Then

$$\mu(E \cap T^{-N-1-s}F) = \mu(E)\,\mu(F)\,(1 + O(\sigma(N)))$$

where the constant in the O-symbol can be choosen independently of E.

Proof: For any cylinder $E = B(k_1,\ldots,k_s)$ define

$$g(x) = \frac{1}{\mu(E)}\, c_E(x)\,\rho(x)$$

We show that $\psi_o(x) = \sum_{b_1,\ldots,b_{s+1}} g(V(b_1,\ldots,b_{s+1})x)\,\Delta(b_1,\ldots,b_{s+1})(x)$

satisfies the assumptions of theorem 8.7.

Clearly $\psi_o(x) = \frac{1}{\mu(E)} \sum_b \rho(V(k_1,\ldots,k_s,b)x)\,\Delta(k_1,\ldots,k_s,b)(x)$

where the sum runs over all b such that $x \in P(k_1,\ldots,k_s,b)$.

We obtain (using theorem 6.4)

$$\psi_o(x) \leq \frac{1}{\mu(E)}\, c_2 \sum_b \Delta(k_1,\ldots,k_s,b)(x) \leq M$$

In the other direction

$$\psi_o(x) \geq \frac{1}{\mu(E)}\, c_1 \sum_b \Delta(k_1,\ldots,k_s,b)(x)$$

where the sum runs over all b such that $B(k_1,\ldots,k_s,b)$ is proper. By lemma 5.1 we obtain

$$\psi_o(x) \geq m > 0$$

with a suitable $m > 0$. Note that m and M can be chosen not dependent on E.

Since $\Delta(k_1,\ldots,k_s,b)(x)$ clearly satisfies a Lipschitz condition we need only

Lemma 1o.3: $\rho(x)$ satisfies a Lipschitz condition on each $B(b_1,\ldots,b_{n-1})$

Proof: Choose a sequence $(\psi_\nu)\,\nu \in N$ according to theorem 8.7 satisfying $a = 1$ (e.g. $\psi_o = 1$). From lemma 8.6 we know

$|\psi_\nu(x) - \psi_\nu(y)| \leq K_1 d(x,y) + K_2\sigma(\nu)$ for $x,y \in B(b_1,\ldots,b_{n-1})$ where K_1 is an absolute constant.

$$|\rho(x) - \rho(y)| \leq |\rho(x) - \psi_\nu(x)| + |\psi_\nu(x) - \psi_\nu(y)| + |\psi_\nu(y) - \rho(y)| \leq$$

$$\leq (b + 2K_2)\,\sigma(\nu) + K_1\, d(x,y)$$

Since ν can be chosen as large as you like we obtain

$$|\rho(x) - \rho(v)| \leq K_1 \, d(x,y)$$

Therefore ψ_o satisfies the assumptions of theorem 8.7 and we obtain

$$|\psi_\nu(x) - \rho(x)| < b\,\sigma(\nu)$$

Note that b can be made not dependent on E. Integration over F gives

$$\left|\int_F \psi_\nu(x) - \mu(F)\right| < b\,\lambda(F)\sigma(\nu)$$

Since $\int_F \psi_\nu(x) = \frac{1}{\mu(E)}\,\mu(E \cap T^{-\nu-s-1} F)$ we have the desired result.

Remark: If all cylinders were proper (which condition holds only for $n = 1$) we could start with

$$\frac{1}{\mu(E)} \Sigma\, \rho(V(k_1,\ldots,k_s)x)\Delta(k_1,\ldots,k_s)(x)$$

to obtain $\mu(E \cap T^{-\nu-s} F) = \mu(E)\mu(F)(1 + O(\sigma(\nu))$

To cover the case of improper cylinders one has to raise s to s+1. One could not use an approximation technique as proposed in Watermann ([2] theorem 6.3) since this approximation is not necessarily uniform.

Theorem 10.4: Let $B(s_1,\ldots,s_t)$ be a fixed cylinder and put

$$A(m,x) = \sum_{\substack{T^k x \,\varepsilon\, B(s_1,\ldots,s_t) \\ 0 \leq k \leq m-1}} 1$$

Then

$$A(m,x) = m\,\mu(B(s_1,\ldots,s_t)) + O(m^{\frac{1}{2}} \log^{\frac{3}{2}+\varepsilon} m)$$

for almost all x.

Proof: Take $E_i = T^{-i} B(s_1,\ldots,s_t)$.

Then

$$\mu(E_i \cap E_{i+j}) = \mu(T^{-i} B(s_1,\ldots,s_t) \cap T^{-i-j} B(s_1,\ldots,s_t)) =$$

$$= \mu(B(s_1,\ldots,s_t) \cap T^{-j} B(s_1,\ldots,s_t)) = \mu(B(s_1,\ldots,s_t))^2 (1+O(\sigma(j-s-1)))$$

for $j \geq s+1$. Clearly the result holds also for $1 \leq j \leq s+1$ (naturally not uniformly in s). Now theorem 10.1 applies giving the result. Here we use corollary 9.7.

With the method used in Schweiger [7] theorem 10.2 can be extended to give a similar but weaker result for the case E an arbitrary intervall. This result allows further limit theorems (e.g. a quantitative formulation of the theorem of "Borel-Bernstein" Schweiger [4],[6]).

Since the function $\sigma(\nu)$ decreases exponentially there may be obtained a number of further limit theorems using the methods developped by Ibragimov [1], Reznik [1], Philipp [1] and others. We refer to the monograph by W.Philipp where further references may be found and to the book by Ibragimov and Linnik [1].

§ 11. Some extensions of Kuzmin's theorem

Theorem 11.1: Let ψ_o be an arbitrary continuous function defined on the unit cube K and define the sequence

$$\psi_{\nu+1}(x) = \sum_k \psi_\nu(V(k)x)\Delta(k)(x)$$

as usually. Then

$$\lim_{\nu\to\infty} \psi_\nu(x) = \rho(x)\int_B \psi_o$$

uniformly in x.

Note: In this theorem we have no convergence rate. With some care one can obtain such a result involving the modulus of continuity.

Proof: If $\psi_o \equiv 0$, then nothing is to be proved. Therefore we may assume $\psi_o \not\equiv 0$. With $A = |\min_{x\varepsilon K} \psi_o(x)|$ we see $\pi_o(x) = A + \psi_o(x) \geq 0$. If we can show the theorem true for all continuous functions $\pi_o \geq 0$ we can easily deduce the general theorem ($\zeta_o(x) \equiv A$) :

$$\lim_{s\to\infty} \pi_s(x) = \rho(x) \int_B \pi_o = \rho(x)(A + \int_B \psi_o)$$

$$\lim_{s\to\infty} \zeta_s(x) = \rho(x) \int_B \zeta_o = \rho(x)\, A$$

Hence $$\lim_{s\to\infty} \psi_s(x) = \rho(x) \int_B \psi_o$$

Here we used linearity of the iteration. Henceforth we assume $\psi_o(x) \geq 0$ and $\psi_o(x) \not\equiv 0$. Since there is at least one $\zeta \in B$ with $\psi_o(\zeta) > 0$ we can assume by continuity:

There is a cylinder $B(a_1,\dots,a_f)$ with the properties:

(i) $\psi_o(x) \geq k > 0$ for $x \in B(a_1,\dots,a_f)$

(ii) $a_1,\dots,a_f, b_1,\dots,b_{n-1}$ is admissible for all admissible $b_1,\dots,b_{n-1}$

Now we conclude as follows:

$$\psi_f(x) = \Sigma\, \psi_o(V(k_1,\dots,k_f)x)\ \Delta(k_1,\dots,k_f)(x) \ge$$
$$\ge \quad \psi_o(V(a_1,\dots,a_f)x)\ \Delta(a_1,\dots,a_f)(x) \qquad \ge \kappa > 0$$

In this way we have proven: There is an index $f \ge 0$ such that $\psi_f(x) \ge \kappa > 0$ for all $x \in B$. We now use:

Lemma 11.2: Let (f_ν), (g_ν) be two sequences obtained by the Kuzmin iteration. If

$$|f_o - g_o| \le \varepsilon \qquad \text{then} \qquad |f_s - g_s| \le \frac{c_2}{c_1}\varepsilon$$

for all $s \ge 0$.

Proof: The assumption $|f_o - g_o| \le \varepsilon$ gives $|f_o(x) - g_o(x)| \le \frac{\varepsilon}{c_1}\ \rho(x)$.

Hence we write

$$f_o(x) = g_o(x) + \eta(x)\ \rho(x) \qquad \text{with } |\eta(x)| \le \frac{\varepsilon}{c_1}$$

Iteration shows

$$f_s(x) = g_s(x) + \Sigma\ \eta(V(k_1,\dots,k_s)x)\Delta(k_1,\dots,k_s)(x).\rho(V(k_1,\dots,k_s)x)$$

$$|f_s(x) - g_s(x)| \le \frac{\varepsilon}{c_1}\ \rho(x) \le \frac{c_2}{c_1}\ \varepsilon$$

We now return to the proof of the theorem. For any cylinder $E = B(k_1,\dots,k_\nu)$ we set $g_o = \frac{1}{\lambda(E)}\ c_E$. Similar to theorem 1o.2 we see

$|g_{s+\nu+1}(x) - \rho(x)| = O(\sigma(s))$ where the O-constant can be choosen uniformly.

To $\varepsilon >$ we can choose a ν such that $|\psi_f(x) - E(\psi_f\|\mathcal{L}^{(\nu)})(x)| \le \varepsilon$ uniformly on B. Clearly $E(\psi_f\|\mathcal{L}^{(\nu)})(x) \ge \kappa > 0$ for all $\nu \ge 0$. Taking $h_o = E(\psi_f\ \|\mathcal{L}^{(\nu)})$ we see from the foregoing remark that

$|h_{s+\nu+1}(x) - a\ \rho\ (x)| = O(\sigma(s))$ where $a = \int_B h_o = \int_B \psi_o$

From lemma 11.2 we obtain

$|\psi_{f+s}\ (x) - a\ \rho(x)| \le \frac{c_2}{c_1}\ \varepsilon + O(\sigma(s))$ which gives the result.

Theorem 11.3: Let P be any absolutely continuous probability measure on B (with respect to Lebesque measure λ), then $\lim_{s\to\infty} P(T^{-s}E) = \mu(E)$ for all Borel sets E.

Proof: Using theorem 1o.2 we see that uniformly (in $B(k_1,\dots,k_s)$)

$\mu(B(k_1,\dots,k_s) \cap T^{-N-1-s}A) = \mu(B(k_1,\dots,k_s))\mu(A)(1 + O(\sigma(N)))$ for any

Borel set A. We can approximate an intervall I by a union of disjoint cylinders

$$K = \bigcup_{B(k_1,\ldots,k_s)\cap K\neq\emptyset} B(k_1,\ldots,k_s)$$

and obtain

$$\mu(K) - \mu(I) = O(\sigma(s))$$

This shows

$$\mu(I\cap T^{-2N}A) = \mu(I)\mu(A) + \mu(A)O(\sigma(N))$$

and therefore

$$\lim_{N\to\infty} \mu(D\cap T^{-N}A) = \mu(D)\mu(A)$$

for any Borel sets D and A. This means

$$\lim_{N\to\infty} \int_{T^{-N}A} c_D(x)\,d\mu = \mu(A) \int_B c_D(x)\,d\mu$$

Since any integrable function f can be approximated in the mean by simple functions (measurable with respect to $\mathcal{L}$) we obtain

$$\lim_{N\to\infty} \int_{T^{-N}A} f(x)\,d\mu(x) = \mu(A) \int_B f(x)\,d\mu$$

Putting $f = \frac{dP}{d\mu}$ we obtain the result.

§ 12. Outer measures

In this chapter we give a brief introduction to the concept of an outer measure. For a more complete account we refer the interested reader Munroe [1] and Rogers [1].

Def.: A function $M^* : \mathcal{P}(\Omega)\to\mathbb{R}\cup\{\infty\}$, where $\mathcal{P}(\Omega)$ denotes the power set of Ω, is called an outer measure if M^* satisfies:

(1) $M^*(\emptyset) = 0$

(2) If $A \subseteq B$, then $M^*(A) \leq M^*(B)$

(3) If $(A_n)_{n\in N}$ is any sequence of subsets of Ω, then

$$M^*(\bigcup_{n\ N} A_n) \leq \sum_{n=1}^{\infty} M^*(A_n)$$

Def.: Let M^* be an outer measure. A set E is called M^*-measurable if

$$M^*(A) = M^*(A\cap E) + M^*(A\setminus E)$$

holds for any set A.

Since clearly $M^*(A) \leq M^*(A \cap E) + M^*(A \setminus E)$ holds by (3) it is sufficient for checking measurability to prove

$$M^*(A) \geq M^*(A \cap E) + M^*(A \setminus E)$$

The first significant result is

Theorem 12.1 (Carathéodory). Let $M^* : \mathcal{P}(\Omega) \to \mathbb{R} \cup \{\infty\}$ be an outer measure. The class $\mathfrak{A}$ consisting of all M^*-measurable sets is a σ-algebra. The restriction of M^* to $\mathfrak{A}$ is a measure (clearly not necessarily a probability measure).

Proof: Clearly, $\emptyset$ is M^*-measurable. The identities

$$A \cap E = A \setminus (\Omega \setminus E)$$

$$A \setminus E = A \cap (\Omega \setminus E)$$

show that if E is M^*-measurable, then so is $\Omega \setminus E$. Now let E_1 and E_2 be M^*-measurable. This implies

$$M^*(A) = M^*(A \cap E_2) + M^*(A \setminus E_2)$$

for any set A.

Take $A \setminus E_1$ to obtain

$$M^*(A \setminus E_1) = M^*((A \setminus E_1) \cap E_2) + M^*((A \setminus E_1) \setminus E_2) =$$

$$= M^*((A \setminus E_1)\ E_2) \cap + M^*(A \setminus (E_1 \cup E_2))$$

Hence

$$M^*(A) = M^*(A \cap E_1) + M^*(A \setminus E_1) =$$

$$= M^*(A \cap E_1) + M^*((A \setminus E_1) \cap E_2) + M^*(A \quad (E_1 \cup E_2))$$

Since $(A \cap E_1) \cup ((A \setminus E_1) \cap E_2) = A \cap (E_1 \cup E_2)$ we have

$M^*(A \cap E_1) + M^*((A \setminus E_1) \cap E_2) \geq M^*(A \cap (E_1 \cup E_2))$ which implies

$M^*(A) \geq M^*(A \cap (E_1 \cup E_2)) + M^*(A \setminus (E_1 \cup E_2))$

This relation is sufficient to insure the M^*-measurability of $E_1 \cup E_2$.

By induction we see: If $E_1, \ldots, E_n$ are M^*-measurable then $S_n = \bigcup_{k=1}^{n} E_k$ is M^*-measurable.

We next show: If the sets $E_1, \ldots, E_n$ are pairwise disjoint, then

$$M^*(A \cap S_n) = \sum_{k=1}^{n} M^*(A \cap E_k)$$

This is true for n=1. We proceed by induction. Taking $A \cap S_n$ as the set checking measurability we see

$$M^*(A\cap S_n) = M^*(A\cap S_n \cap S_{n-1}) + M^*((A\cap S_n) \setminus S_{n-1}) =$$

$$= M^*(A\cap S_{n-1}) + M^*(A\cap E_n) = \sum_{k=1}^{n-1} M^*(A\cap E_k) + M^*(A\cap E_n) = \sum_{k=1}^{n} M^*(A\cap E_k)$$

For any sequence $(E_k)_{k\in N}$ of disjoint M^*-measurable sets we denote

$S = \bigcup_{k=1}^{\infty} E_k$. Clearly $A\cap S_n \subseteq A\cap S$ for every $n \geq 1$.

Hence

$$M^*(A\cap S) \geq M^*(A\cap S_n) = \sum_{k=1}^{n} M^*(A\cap E_k)$$

This shows

$$M^*(A\cap S) \geq \sum_{k=1}^{\infty} M^*(A\cap E_k)$$

Since $\bigcup_{k=1}^{\infty} (A\cap E_k) = A\cap S$ the inequality in the other direction is true by the very definition of an outer measure.

Therefore we obtain $M^*(A\cap S) = \sum_{k=1}^{\infty} M^*(A\cap E_k)$

We are now in position to show S is M^*-measurable. For any set A we have

$$M^*(A) = M^*(A\cap S_n) + M^*(A\setminus S_n) \geq \sum_{k=1}^{n} M^*(A\cap E_k) + M^*(A\setminus S)$$

Letting $n \to \infty$ we see $M^*(A) \geq M^*(A\cap S) + M^*(A\setminus S)$.

Since any countable union of sets is representable as a countable union of disjoint sets we have proven that $\mathfrak{A}$ is a σ-algebra and the restricti of M^* to $\mathfrak{A}$ is a measure.

Remark: The measure obtained in this way is complete in the following sense. If $M^*(E) = 0$, then E is M^*-measurable. This can be seen very easily as follows:
From $A\cap E \subseteq E$ and $A\setminus E \subseteq A$ we have

$$M^*(A\cap E) + M^*(A\setminus E) \leq M^*(E) + M^*(A) = M^*(A).$$

Note that the property of completeness is more a property of the σ-algebra involved. Lebesque measure, for instance, is not complete on the σ-algebra of Borel sets, but complete on the σ-algebra of all Lebesque measurable sets.

Def.: A function $\tau : \mathcal{C} \to \mathbb{R} \cup \{+\infty\}$ defined on a class $\mathcal{C}$ of subsets of is called a pre-measure, if

(1) $\emptyset \in \mathcal{C}$

(2) $\tau(\emptyset) = 0$

(3) $0 \leq \tau(C)$ for all $C \in \mathcal{C}$

Def.: Given a pre-measure τ on a class $\mathcal{C}$ we define a set function

$$M^*(A) = M^*(\tau, \mathcal{C}; A) = \inf \{ \sum_{n=1}^{\infty} \tau(C_n) \mid A \subseteq \bigcup_{n=1}^{\infty} C_n, C_n \in \mathcal{C} \}$$

Remark: We put $\inf \emptyset = +\infty$

A sequence $(C_i)_{i \in \mathbb{N}}$ with $A \subseteq \bigcup_{i=1}^{\infty} C_i$ will be called a $\mathcal{C}$-covering of A.

Theorem 12.2: Given a pre-measure τ on $\mathcal{C}$, the set function M^* is an outer measure.

Proof: Since $\emptyset \in \mathcal{C}$, we have $M^*(\emptyset) = 0$. If $A \subseteq B$, any $\mathcal{C}$-covering of B is a $\mathcal{C}$-covering of A and therefore

$$M^*(A) \leq M^*(B)$$

Let $(A_i)_{i \in \mathbb{N}}$ be any sequence of subsets of Ω. Given $\varepsilon > 0$ there are $\mathcal{C}$-coverings $(C_{ik})_{k \in \mathbb{N}}$ of A_i, $i \in \mathbb{N}$, with

$$\sum_{k=1}^{\infty} \tau(C_{ik}) \leq M^*(A_i) + \frac{\varepsilon}{2^i}$$

(with the usual convention if $M^*(A_i)$ is infinite). Clearly $(C_{ik})_{i,k \in \mathbb{N}}$ is a $\mathcal{C}$-covering of $\bigcup_{i=1}^{\infty} A_i$, hence

$$M^*(\bigcup_{i=1}^{\infty} A_i) \leq \sum_{i=1}^{\infty} M^*(A_i) + \varepsilon$$

Since $\varepsilon > 0$ was arbitrary the result follows.

Proposition 12.3: Let τ be a pre-measure on a class $\mathcal{C}$. If $\mathcal{D}$ is a subclass of $\mathcal{C}$ containing $\emptyset$, then the restriction of τ to $\mathcal{D}$ is a pre-measure and $M^*(\tau, \mathcal{C}; A) \leq M^*(\tau, \mathcal{D}; A)$ for any set A.

Proof: It is sufficient to note that any $\mathcal{D}$-covering is a $\mathcal{C}$-covering.

Theorem 12.4: Let τ be a pre-measure on $\mathcal{C}$ and $\mathcal{D} \subseteq \mathcal{C}$ a subclass containing $\emptyset$. If $M^*(\tau, \mathcal{D}; C) \leq \tau(C)$ for all $C \in \mathcal{C}$, then

$$M^*(\tau, \mathcal{D}; A) = M^*(\tau, \mathcal{C}; A) \quad \text{for all } A \subseteq \Omega.$$

Proof: The assumption shows that given $\varepsilon > 0$ and a $\mathcal{C}$-covering $(C_i)_{i \in \mathbb{N}}$

of A we may choose $(D_{ik})_{k\in\mathbb{N}}$, $i\in\mathbb{N}$, with

$$\sum_{k=1}^{\infty} \tau(D_{ik}) \leq \tau(C_i) + \frac{\varepsilon}{2^i}$$

Since $(D_{ik})_{i,k\in\mathbb{N}}$ is a $\mathcal{D}$-covering we obtain

$$M^*(\tau,\mathcal{D};A) \leq \sum_{i=1}^{\infty} \tau(C_i) + \varepsilon$$

for every $\mathcal{C}$-covering. This shows

$$M^*(\tau,\mathcal{D};A) \leq M^*(\tau,\mathcal{C};A)$$

Proposition 12.3 gives the result.

Example: Take $\Omega = K$, the n-dimensional unit cube, $\tau = \lambda$, the n-dimensional Lebesque measure and $\mathcal{C} = \mathcal{L}$, the class of all Borel sets. One can show that

$$M^*(\lambda,\mathcal{L};A) = \lambda(A)$$

for all Lebesque measurable sets A.

Theorem 12.4 shows that one can restrict the coverings to various subclasses: the open sets, the closed sets, the class of all intervalls, the class of all open spheres. It is shown in treatises of measure theory that the crucial assumption of theorem 12.4 is satisfied.

§ 13. Hausdorff measures

Theorem 13.1: Let $\mathcal{C}$ be a covering class with pre-measure τ defined on it. For any set of subclasses $S = (\mathcal{C}_\alpha)_{\alpha\in A}$, $\mathcal{C}_\alpha \subseteq \mathcal{C}$, define

$$\mu^*(\tau, S;B) = \sup \{M^*(\tau,\mathcal{C}_\alpha;B) \mid \alpha\in A\}$$

(the supremum may possibly be $+\infty$). Then μ^* is an outer measure.

Proof: Clearly $\mu^*(\tau,S;B) \geq 0$ for all subsets $B \subseteq \Omega$. Since $M^*(\tau,\mathcal{C}_\alpha;\emptyset) = 0$ for all $\mathcal{C}_\alpha \in S$, we have $\mu^*(\tau,S;\emptyset) = 0$. Let $B \subseteq D$. Then $M^*(\tau,\mathcal{C}_\alpha;B) \leq M^*(\tau,\mathcal{C}_\alpha;D) \leq \mu^*(\tau,S;D)$ hence $\mu^*(\tau,S;B) \leq \mu^*(\tau,S;D)$

For any sequence $(B_n)_{n\in\mathbb{N}}$ of subsets of Ω, we obtain

$$M^*(\tau,\mathcal{C}_\alpha;\bigcup_{n=1}^{\infty} B_n) \leq \sum_{n=1}^{\infty} M^*(\tau,\mathcal{C}_\alpha;B_n) \leq \sum_{n=1}^{\infty} \mu^*(\tau,S;B_n)$$

Therefore we obtain $\mu^*(\tau,S;\bigcup_{n=1}^{\infty} B_n) \leq \sum_{n=1}^{\infty} \mu^*(\tau,S;B_n)$

Theorem 13.2: Under the conditions of theorem 13.1 let $\mathcal{D}$ be a subset of $\mathcal{C}$ containing $\emptyset$. The set $S' = (\mathcal{C}_\alpha \cap \mathcal{D})_{\alpha \in A}$ satisfies $\mathcal{C}_\alpha \cap \mathcal{D} \subseteq \mathcal{D}$ for all $\alpha \in A$ and $\mu^*(\tau,S';B) \geq \mu^*(\tau,S;B)$

Proof: From proposition 12.3 we see

$$M^*(\tau,\mathcal{C}_\alpha \cap \mathcal{D};B) \geq M^*(\tau,\mathcal{C}_\alpha;B)$$

From this the result is obvious.

Theorem 13.3: If furthermore $M^*(\tau,\mathcal{C}_\alpha \cap \mathcal{D};C) \leq \tau(C)$ for all $C \in \mathcal{C}_\alpha$, then $\mu^*(\tau,S';B) = \mu^*(\tau,S;B)$

Proof: Theorem 12.4 implies that $M^*(\tau,\mathcal{C}_\alpha \cap \mathcal{D};B) = M^*(\tau,\mathcal{C}_\alpha;B)$ holds under this condition.

Example: Take $\Omega = K$, the n-dimensional unit cube, $\tau = \lambda$ and $\mathcal{C} = \mathcal{L}$. Now let $A = \{\alpha \in \mathbb{R} \mid \alpha > 0\}$ and define

$$\mathcal{L}_\alpha := \{C \in \mathcal{L} \mid \text{diam } C \leq \alpha\}$$

Since in this case $M^*(\lambda,\mathcal{L}_\alpha;D) = M^*(\lambda,\mathcal{L};D) = \lambda(D)$ for all measurable sets D, clearly $\mu^*(\lambda,S;D) = \lambda(D)$

One can show by suitable approximation that the condition of theorem 13.3 is satisfied for various subclasses $\mathcal{D}$: the open sets, the closed sets, the set of intervalls, the set of spheres and so on.

We now come to the most important examples:

Def.: Let Ω be a set. A function $d: \Omega \times \Omega \to \mathbb{R}$ is called a semimetric (or pseudometric), if

(1) $d(x,y) \geq 0$ and $d(x,x) = 0$

(2) $d(x,y) = d(y,x)$

(3) $d(x,y) + d(y,z) \geq d(x,z)$

A semimetric is called a metric, if additionally $d(x,y) = 0$ implies $x = y$.

The pair (Ω,d) is called a semimetric (or pseudometric) space.

Def.: The diameter of any set $A \subseteq \Omega$ is defined as

$$\text{diam } A = \sup\{d(x,y) \mid x,y \in \Omega\}$$

Proposition 13.4: Let (Ω,d) be a semimetric space and $\mathcal{C} \subseteq P(\Omega)$ with $\emptyset \in \mathcal{C}$. For any real $\gamma \geq 0$ the function $h(\gamma,\cdot) : \mathcal{C} \to \mathbb{R} \cup \{+\infty\}$ defined as $h(\gamma,C) = (\text{diam } C)^\gamma$ is a pre-measure.

We now take $S = (\mathcal{C}_\alpha)_{\alpha \in A}$, $A = \{\alpha \in \mathbb{R} \mid \alpha > 0\}$, defining $\mathcal{C}_\alpha = \{C \in \mathcal{C} \mid \text{diam } C \leq \alpha\}$. If $\mathcal{C}_\alpha = \emptyset$, then $M^*(h(\gamma,\cdot), \mathcal{C}_\alpha; D) = \infty$. The resulting outer measure $H^*(\gamma, \mathcal{C}; D) = \sup \{M^*(h(\gamma,\cdot), \mathcal{C}_\alpha; D) \mid \alpha \in A\}$ is called the γ-dimensional Hausdorff measure with respect to the co - vering class $\mathcal{C}$. Since $\mathcal{C}_\alpha \subseteq \mathcal{C}_\beta$ if $\alpha \leq \beta$ clearly $M^*(h(\gamma,\cdot), \mathcal{C}_\beta; D) \leq$ $\leq M^*(h(\gamma,\cdot), \mathcal{C}_\alpha; D)$ in this case and the supremum is the limit of an increasing function. In a briefly compact version we may write

$$H^*(\gamma, \mathcal{C}; D) = \lim_{\alpha \to 0} \inf \sum_{n=1}^{\infty} (\text{diam } C_n)^\gamma$$

where the infimum is extended over all $\mathcal{C}$-coverings of D with diam $C_n \leq \alpha$. There is a generalisation of γ-dimensional Hausdorff measure taking pre-measure of the form $\tau(C) = h(\text{diam } C)$, where $h(t)$ is monotonic increasing for $t \geq 0$, positive for $t > 0$ and continuous on the right for all $t \geq 0$. Various authors assume $h(0) = 0$ and h being concave (see e.g. Rogers [1], Federer [1], Kahane-Salem [1]).

The most important case is taking $\mathcal{C} = P(\Omega)$, $\mathcal{C} = \mathcal{B}$ (the σ-algebra of Borel sets in Ω, that is the smallest σ-algebra containing the open or equally the closed sets of the topology induced by the semimetric) or various subsets of $\mathcal{B}$ (open sets, closed sets, open spheres and so on). Various authors use to define Hausdorff measure by various classes of subsets. Fortunately one can show that in most cases the differently defined measures are very intimately related. The most complete account on these questions may be found in the remarkable book by Federer [1].

<u>Proposition 13.5</u>: For any covering class $\mathcal{C}$ denote with $\overline{\mathcal{C}}$ the class obtained from $\mathcal{C}$ by taking closures. Then $H^*(\gamma, \mathcal{C}; D) = H^*(\gamma, \overline{\mathcal{C}}, D)$ for all $D \subseteq \Omega$.

<u>Proof</u>: For any set $M \subseteq \Omega$ clearly diam M = diam $\overline{M}$ ($\overline{M}$ is the closure of M with respect to the topology induced by the semimetric d). Note that a corresponding result for interiors does not hold generally.

<u>Example</u>: Take $\Omega = \mathbb{R}^2$, the 2-dimensional Euclidean space with its usual metric. $\mathcal{C}$ may be equal the set of all line segments, then the set $\mathcal{C}^\circ$ of all interiors reduces to $\{\emptyset\}$. Hence $H^*(\gamma, \mathcal{C}^\circ; D) = +\infty$ for all sets $D \neq \emptyset$. Clearly there are sets with $H^*(\gamma, \mathcal{C}; D)$ finite.

It is obvious that diam C° = diam C for all $C \in \mathcal{C}$ (C° be the interior of C) is sufficient to insure

$$H^*(\gamma, \mathcal{C}; D) = H^*(\gamma, \mathcal{C}^\circ; D).$$

Theorem 13.6: If $\mathcal{T}$ denotes the set of all open spheres of a semimetric space (Ω,d), then

$$H^*(\gamma,\mathcal{P}(\Omega);D) \leq H^*(\gamma,\mathcal{T};D) \leq 2^\gamma H^*(\gamma,\mathcal{P}(\Omega);D)$$

Proof: Since every set of diameter $\leq \alpha$ is contained in an open sphere of diameter $\leq 3\alpha$, we see

$$M^*(h(\gamma,\cdot),\mathcal{T}_{3\alpha};D) \leq 2^\gamma M^*(h(\gamma,\cdot),\mathcal{P}(\Omega)_\alpha;D)$$

Letting $\alpha \to 0$ we obtain the result.

Remark 1: Since the Borel sets are between $\mathcal{T}$ and $\mathcal{P}(\Omega)$, a similar result is valid for them.

Remark 2: In n-dimensional Euclidean space one can replace 2^γ by

$\left(\frac{2n}{n+1}\right)^{\frac{\gamma}{2}}$ (see Federer [1], p.201).

Remark 3: In n-dimensional space one can prove (Federer [1], Rogers [1]): For $\gamma = n$ there is a constant κ_n with $H^*(n,\mathcal{P}(\Omega);D) = \kappa_n\lambda(D)$ for every Lebesque measurable D. It can be shown

$$\kappa_n = \frac{2^n\,\Gamma(\frac{n}{2}+1)}{\Gamma(\frac{1}{2})^n}$$

Theorem 13.7: For any subset $A \subseteq \Omega$

$$H^*(\gamma,\mathcal{P}(A);A) = H^*(\gamma,\mathcal{P}(\Omega);A)$$

Proof: Since $\mathcal{P}(A) \subseteq \mathcal{P}(\Omega)$ clearly $H^*(\gamma,\mathcal{P}(A);A) \geq H^*(\gamma,\mathcal{P}(\Omega);A)$

On the other hand, let $A \subseteq \bigcup_{i=1}^\infty C_i$, then $A \bigcup_{i=1}^\infty (C_i \cap A)$.

Then $\sum_{i=1}^\infty (\text{diam } C_i)^\gamma \geq \sum_{i=1}^\infty (\text{diam } C_i \cap A)^\gamma$ and therefore

$$H^*(\gamma,\mathcal{P}(\Omega);A) \geq H^*(\gamma,\mathcal{P}(A);A)$$

§ 14. Hausdorff dimension

In this chapter we are concerned with a semimetric space (Ω,d), the premeasure $h(\gamma,C) = (\text{diam } C)^\gamma$, defined on a covering class $\mathcal{C}$ and the set $S = (\mathcal{C}_\alpha)_{\alpha>0}$, where $\mathcal{C}_\alpha = \{C \in \mathcal{C} \mid \text{diam } C \leq \alpha\}$.

Theorem 14.1:

(1) If $H^*(\gamma,\mathcal{C};D) < \infty$, then $H^*(\beta,\mathcal{C};D) = 0$ for all $\beta > \gamma$.

(2) If $H^*(\gamma,\mathcal{L};D) > 0$, then $H^*(\delta,\mathcal{L};D) = +\infty$ for all $\delta < \gamma$.

Proof: It is easy to see that (1) and (2) are equivalent. We prove (1). To every $\varepsilon > 0$ we can choose $t_0 > 0$ such that

$$t^\beta \leq \varepsilon\, t^\gamma \quad , \quad 0 < t \leq t_0$$

Let $D \subseteq \bigcup_{k=1}^{\infty} C_k$ where $C_k \in \mathcal{L}_\alpha$ and $\alpha \leq t_0$, then

$$\sum_{k=1}^{\infty} (\operatorname{diam} C_k)^\beta \leq \varepsilon \sum_{k=1}^{\infty} (\operatorname{diam} C_k)^\gamma$$

Hence $M^*(h(\beta,\cdot),\ \mathcal{L}_\alpha;D) \leq \varepsilon\, M^*(h(\gamma,\cdot),\mathcal{L}_\alpha;D)$

Let $\alpha \to 0$ then we see $H^*(\beta,\mathcal{L};D) \leq \varepsilon\, H^*(\gamma,\mathcal{L};D)$

Since $\varepsilon > 0$ was arbitrary we have the result.

Def.: $\dim(\mathcal{L};D) = \sup\{\gamma \mid H^*(\gamma,\mathcal{L};D) = \infty\} = \inf\{\gamma \mid H^*(\gamma,\mathcal{L};D) = 0\}$

The number (possibly $+\infty$) $\dim(\mathcal{L};D)$ is called the Hausdorff dimension with respect to the covering class $\mathcal{L}$.

Proposition 14.2: If $0 < H^*(\gamma,\mathcal{L};D) < \infty$, then $\dim(\mathcal{L};D) = \gamma$.

Proof: Obvious by the very definition of the dimension.

Proposition 14.3: The set function $\dim(\mathcal{L};\cdot)$ is an outer measure.

Proof: Since $H^*(\gamma,\mathcal{L};\emptyset) = 0$, clearly $\dim(\mathcal{L};\emptyset) = 0$. Furthermore $\dim(\mathcal{L};D) \leq \dim(\mathcal{L};E)$, if $D \subseteq E$. Last we prove

Lemma 14.4: For any sequence $(D_k)_{k\in\mathbb{N}}$

$$\dim(\mathcal{L};\ \bigcup_{k=1}^{\infty} D_k) = \sup\,\{\dim(\mathcal{L};D_k)\}\ .$$

Proof: Since $D_n \subseteq \bigcup_{k=1}^{\infty} D_k$, we see $\dim(\mathcal{L};D_n) \leq \dim(\mathcal{L};\ \bigcup_{k=1}^{\infty} D_k)$

and $\sup\,\{\dim(\mathcal{L};D_n)\} \leq \dim(\mathcal{L};\ \bigcup_{k=1}^{\infty} D_k)$

By theorem 13.1. the set function $H^*(\gamma,\mathcal{L};\cdot)$ is an outer measure.

Therefore $H^*(\gamma,\mathcal{L};\ \bigcup_{k=1}^{\infty} D_k) \leq \sum_{k=1}^{\infty} H^*(\gamma,\mathcal{L};D_k)$

In the case $\sup\{\dim(\mathcal{L};D_k)\} = \infty$ there is nothing to prove. In the other case take $\gamma > \sup\{\dim(\mathcal{L};D_k)\}$. Hence $\gamma > \dim(\mathcal{L};D_k)$ and $H^*(\gamma,\mathcal{L};D_k) = 0$. This shows $H^*(\gamma,\mathcal{L};\ \bigcup_{k=1}^{\infty} D_k) = 0$ and $\dim(\mathcal{L};\bigcup_{k=1}^{\infty} D_k) \leq \gamma$.

We now wish to compare different covering classes $\mathcal{L}$ and $\mathcal{D}$.

Lemma 14.5: Suppose that there is a constant $c > 0$ such that

$H^*(\gamma,\mathcal{L};E) \leq c\, H^*(\gamma,\mathcal{D};E)$ then $\dim(\mathcal{D};E) \geq \dim(\mathcal{L};E)$

Proof: If $H^*(\gamma,\mathcal{D};E) = 0$ then $H^*(\gamma,\mathcal{L};E) = 0$.

Corollary 1: If $\mathcal{D} \subseteq \mathcal{L}$ then one can take $c = 1$ to obtain $\dim(\mathcal{D};E) \geq \dim(\mathcal{L};E)$

Corollary 2: Let $\mathcal{P}(\Omega)$ be the power of the semimetric space (Ω,d), $\mathcal{G}$ be the set of all open sets, $\mathcal{T}$ the set of all open spheres, then

$$\dim(\mathcal{P}(\Omega);E) = \dim(\mathcal{G};E) = \dim(\mathcal{T};E)$$

This follows from theorem 13.6. Since the set $\mathcal{F}$ of all closed sets contains all closures of open sets $\dim(\mathcal{P}(\Omega);E) = \dim(\mathcal{F};E)$. Henceforth we will write $\dim E$ for $\dim(\mathcal{P}(\Omega);E)$. One of the major problems of dimension theory is to show $\dim E = \dim(\mathcal{L};E)$ for various covering classes $\mathcal{L}$, usually smaller than $\mathcal{G}$ or $\mathcal{F}$. Note that corollary 1 implies $\dim E \leq \dim(\mathcal{L};E)$ for all covering classes $\mathcal{L}$. The set function dim is sometimes referred as Hausdorff-Besicovitch dimension or fractional dimension.

Lemma 14.6: If $\Omega = \mathbb{R}^n$ equipped with its usual Euclidean metric, then

$$\dim E \leq n$$

Proof: From remark 3 given at the end of theorem 13.6 we recall $H^*(n,\mathcal{P}(\Omega);E) = O\,\kappa_n\,\lambda(E)$ for every Lebesque measurable set E. Hence $\dim E \leq n$ for every set E of finite Lebesque measure. Since every set can be covered by at most countably many sets of finite measure the result follows from lemma 14.4 .

§ 15. Billingsley dimension

We now return to a probability space $(\Omega,\mathcal{F},P)$. A function $f : \Omega \to R^n$ is called measurable, if the counterimage of any n-dimensional Borel set is measurable $\mathcal{F}$.

Def.: A sequence $(f_k)_{k\in\mathbb{N}}$ of measurable functions $f_k : \Omega \to \mathbb{R}^n$ is called a stochastic process.

In the sequel we are only concerned with stochastic processes with "discrete state space" that means the set of values taken by the sequence $(f_k)_{k\in\mathbb{N}}$ is a discrete set. Hence it may be assumed as a subset of $\mathbb{N}_o^{\,n}$.

Example: $(\Omega, \mathcal{F}, P) = (B, \mathcal{L}, \lambda)$.
Take $f_s = k_s$, the s-th digit function of Jacobi algorithm. In the following the discrete state space will be denoted by S.

Def.: A cylinder of order m is a set

$$Z(s_1,\ldots,s_m) = \{\omega \in \Omega \mid f_k(\omega) = s_k, \; 1 \le k \le m\}$$

for the sequence of digit functions this concept coincides with that given in § 2.

We now use a device given by Wegmann [1] to introduce Billingsley dimension as a special case of Hausdorff dimension. Most of the following material is adapted from Wegmann [1].

We later briefly refer Billingsley's original approach (Billingsley [2], [3]).

The stochastic process $(f_k)_{k \in \mathbb{N}}$ on the probability space $(\Omega, \mathcal{F}, P)$ now satisfies the following two conditions:

(A) $\lim_{k\to\infty} P(Z(s_1,\ldots,s_k)) = 0$ for all sequences $(s_k)_{k\in\mathbb{N}}$, $s_k \in S$.

Under the assumption of condition (A) we give:

Def.: $d^*(x,y) = \inf\{P(Z(s_1,\ldots,s_k)) \mid x,y \in Z(s_1,\ldots,s_k)\}$. In the sequel all "metric" properties of (Ω, d^*) will be starred.

Theorem 15.1 (Wegmann): (Ω, d^*) is a semimetric space.

Proof: Clearly $d^*(x,y) = d^*(y,x) \ge 0$.
Condition (A) insures $d^*(x,x) = 0$.

We further show $d^*(x,z) \le \max(d^*(x,y), d^*(y,z))$
If $d^*(x,z) = 0$, there is nothing to prove. If $d^*(x,z) > 0$, there is a cylinder $Z(s_1,\ldots,s_k, s_{k+1})$ such that $x \in Z(s_1,\ldots,s_{k+1})$ and $z \in Z(s_1,\ldots,s_k)$, but $z \notin Z(s_1,\ldots,s_k, s_{k+1})$. Since different cylinders of the same order are disjoint $d^*(x,z) = P(Z(s_1,\ldots,s_k))$.
Suppose that $y \in Z(s_1,\ldots,s_{k+1})$, then $d^*(y,z) = P(Z(s_1,\ldots,s_k)) = d^*(x,z)$.
In the other case $d^*(x,y) \ge P(Z(s_1,\ldots,s_k)) = d^*(x,z)$.

Remark: Taking $(\Omega, \mathcal{F}, P) = (B, \mathcal{L}, \lambda)$ and the stochastic process equal the sequence of the digit functions condition (A) is satisfied. Therefore Jacobi algorithm induces a new semimetric on B. This semimetric will be denoted as d^*_J to distinguish it from (induced) Euclidean metric d_E. The sharper form of the triangel inequality as proven in theorem 15.1 shows that these two metrics are in fact extremely different from eachother. The uniqueness of the Jacobi algorithm shows that d^*_J is in fact a metric.

Now we denote the set of all cylinders with $\mathfrak{Z}$. Note that $\mathfrak{Z} = \bigcup_{s=1}^{\infty} \mathfrak{Z}^{(s)}$, $\mathfrak{Z}^{(s)}$ is the set of all cylinders of order s, is different from $\bigvee_{s=1}^{\infty} \mathfrak{Z}^{(s)}$, the σ-algebra generated by $\mathfrak{Z}$.

Since the intersection of two cylinders is itself a cylinder or the empty set and every $\omega \in \Omega$ is contained in at least one cylinder, the class $\mathfrak{Z}$ generates a topology $\mathcal{T}$ on Ω with base $\mathfrak{Z}$ (see e.g.Gaal [1], p.33).

We now assume

(B) $P(Z(s_1,\dots,s_k)) > 0$ for every cylinder. Then we can prove

Theorem 15.2: The topology $\mathcal{T}$ coincides with the topology induced by the semimetric d^*.

Proof: Given a cylinder $B(s_1,\dots,s_k)$ and $x \in B(s_1,\dots,s_k)$ then the open sphere $\{y \mid d^*(x,y) < P(B(s_1,\dots,s_k))\}$ is a subset of $B(s_1,\dots,s_k)$. Conversely, let us be given an open sphere $\{y \mid d^*(x,y) < r\}$, $r > 0$. Since the sequence of cylinders $B(s_1,\dots,s_k) \geq B(s_1,\dots,s_k, s_{k+1})$ containing x is unique and $\lim_{k\to\infty} P(B(s_1,\dots,s_k)) = 0$, there is a cylinder $B(s_1,\dots,s_k)$ with $x \in B(s_1,\dots,s_k)$ and $P(B(s_1,\dots,s_k)) < r$. Therefore $B(s_1,\dots,s_k) \subseteq \{y \mid d^*(x,y) < r\}$.

In the semimetric space (Ω,d^*) we can define γ-dimensional measures. We first show:

Theorem 15.3: $H^*(\gamma,\mathfrak{Z};D) = H^*(\gamma,\mathcal{P}(\Omega);D)$

Proof: We have only to show

$$H^*(\gamma,\mathcal{P}(\Omega);D) \geq H^*(\gamma,\mathfrak{Z};D)$$

It is enough to show the following: Given any set C with $\operatorname{diam}^* C > 0$ there is a cylinder Z with $\operatorname{diam}^* C = \operatorname{diam}^* Z$ and $C \subseteq Z$. Take $Z = Z(s_1,\dots,s_k)$ the cylinder with $C \subseteq Z(s_1,\dots,s_k)$ but $C \not\subseteq Z(s_1,\dots,s_k,s_{k+1})$ for at least one s_{k+1}. Such a cylinder exists (possibly $k = 0$ and $Z = \Omega$). Since $\lim_{k\to\infty} P(Z(s_1,\dots,s_k)) = 0$ and $\operatorname{diam}^* C > 0$, we choose $x \in C$ and the sequence of cylinders $Z(s_1,\dots,s_k)$

containing x will give such a cylinder. Clearly $\operatorname{diam}^* C \leq$
$\leq \operatorname{diam}^* Z(s_1,\dots,s_k) = P(Z(s_1,\dots,s_k))$.

On the other hand, given $x \in Z(s_1,\dots,s_k,s_{k+1}) \cap C$ there is a $y \in C$ satisfying $y \notin Z(s_1,\dots,s_{k+1})$. Therefore $d^*(x,y) \geq \operatorname{diam} Z(s_1,\dots,s_k)$ and the result comes out.

Note that theorem 15.3 is only true for a semimetric space the semimetric of which is induced by the measure of cylinders.

These new γ-dimensional measures define a dimension on Ω according to § 14:

Def.: The Billingsley dimension of a set D is defined as

$$\operatorname{bim} D = \dim^*(\mathfrak{Z};D) = \sup\{\gamma \mid H^*(\gamma,\mathfrak{Z};D) = \infty\} = \inf\{\gamma \mid H^*(\gamma,\mathfrak{Z};D) = 0\}$$

We now refer briefly Billingsley's original approach: He defines

$$L_\rho(M,\gamma) = \inf\{\sum_{i=1}^{\infty} P(Z_i)^\gamma \mid M \subseteq \bigcup_{i=1}^{\infty} Z_i \,,\ P(Z_i) < \rho \text{ and } Z_i \in \mathfrak{Z}\}$$

As ρ decreases to 0, $L_\rho(M,\gamma)$ increases to a (possibly infinite) limit $L(M,\gamma)$. It is easy to see that $L(M,\gamma) = H^*(\gamma,\mathfrak{Z};M)$ because $P(Z_i) < \rho$ is equivalent to $\operatorname{diam}^* Z_i < \rho$ for any cylinder Z_i. Theorem 15.3 shows th in this case the restriction to coverings by cylinders is not essential

§ 16. Comparison theorems

Given a set Ω there may be given two semimetrics d_1 and d_2 which make Ω a semimetric space. It is an important problem to compare the di - mensions given in this way. The most common examples for this situatior are:

(1) A measure space $(\Omega,\mathcal{F})$ is equipped with two probability measures P_1 and P_2. A given stochastic process $(f_n)_{n \in z}$ therefore induces two semimetrics d_1 and d_2 which both will lead to bim_1 and bim_2.

(2) A set Ω is both a measure space $(\Omega,\mathcal{F},P)$ and a semimetric space (Ω,d). A stochastic process $(f_n)_{n \in z}$ gives a set of cylinders $\mathfrak{Z}$ and a new semimetric d^*. The problem of comparing dim and bim (induced by d^*) can be divided into two steps:

(a) Compare $\operatorname{bim} = \dim^*(\mathfrak{Z};D)$ with $\dim(\mathfrak{Z};D)$

(b) Compare $\dim(\mathfrak{Z};D)$ with dim D.

In several problems step (a) becomes easy: Take $\Omega = [0,1]$ and $P = \lambda$. If all cylinders are intervalls, then clearly

$$\lambda(Z) = \operatorname{diam}^* Z = \operatorname{diam} Z$$

(the starred diam belongs to d^*) and $\dim^*(\mathfrak{Z};D) = \dim(\mathfrak{Z};D)$. Unfortunately for $\Omega = K$, the n-dimensional unit cube, this does not hold for $n \geq 2$. If $P = \lambda$ one has to compare $\operatorname{diam}^* Z = \lambda(Z)$ with diam Z. In this case step (a) is almost unexplored.

It is worthwhile to note that even in the case $\operatorname{diam}^* Z = \operatorname{diam} Z$ step (b) is not trivial. In fact theorem 15.3 holds only for the metric d^* and not for the metric d. Note, that $\operatorname{diam}^* Z = \operatorname{diam} Z$ does not imply that $d^* = d$.

The only cases, where step (b) is handled, assume that $\Omega = [0,1]$ and all cylinders should be intervals (Billingsley [1] , Kinney-Pitcher [1] Wegmann [1]).

We now look at a set Ω equipped with two semimetrics d_1 and d_2. As usual, we define an open sphere as

$$K_i(x,r) = \{y \mid d_i(x,y) < r\} , \quad i = 1,2 .$$

Then we prove

<u>Theorem 16.1 (Wegmann [1]):</u>

If $\quad \liminf\limits_{r \to 0} \dfrac{\log \operatorname{diam}_2 (K_1(x,r))}{\log \operatorname{diam}_1 (K_1(x,r))} \geq \delta \qquad$ for all $x \in M$,

then $\quad \dim_1 M \geq \delta \dim_2 M$

<u>Remark 1</u>: We adopt the conventions:

$$\frac{\log \xi}{\log 0} = 0 \qquad , \; 0 < \xi \leq 1$$

$$\frac{\log 0}{\log \eta} = \infty \qquad , \; 0 < \eta \leq 1$$

$$\frac{\log 0}{\log 0} = \frac{\log 1}{\log 1} = 1$$

$$\frac{\log \xi}{\log 1} = \infty \qquad , \; 0 \leq \xi < 1$$

<u>Proof</u>: For a set $M \subseteq \Omega$ we define

$$d_2(g,M) = \inf\{\sum_{i=1}^{g} \operatorname{diam}_2 A_i \mid M \subseteq \bigcup_{i=1}^{g} A_i\}$$

where $g \geq 1$ is any fixed integer. Clearly, $d_2(g,M)$ is a nonincreasing function as g increases and $d_2(1,M) = \mathrm{diam}_2 M$. We will prove a considerably sharper theorem which is also due to Wegmann:

Theorem 16.2:

If $$\liminf_{r \to 0} \frac{\log d_2 (g,K_1(x,r))}{\log \mathrm{diam}_1 K_1(x,r)} \geq \delta \quad \text{for all } x \in M ,$$

then $\dim_1 M \geq \delta \dim_2 M$.

Proof: For $\delta \leq 0$ the assertion is clearly true. Hence assume $\delta > 0$. We put $\beta = \dim_2 M$. If $\beta = 0$ there is nothing to prove, hence we also assume $\beta > 0$. Now we choose an ε with $0 < \varepsilon < \min(\delta,\beta)$

For any $x \in M$ we have

$$\frac{\log d_2(g, K_1(x,r))}{\log \mathrm{diam}_1 K_1(x,r)} > \delta - \varepsilon \quad \text{for } r \leq r(\varepsilon). \text{ We define}$$

$$M_n = \{x \in M \mid d_2(g,K_1(x,r)) \leq [\mathrm{diam}_1 K_1(x,r)]^{\delta-\varepsilon} \quad \text{for all} \quad r < \tfrac{1}{n} \}$$

Then $\bigcup_{n=1}^{\infty} M_n = M$ and by lemma 14.4 we have $\dim_i M = \lim_{n\to\infty} \dim_i M_n$ for both $i = 1,2$.

Since $\dim_2 M = \beta$, clearly $\dim_2 M_n > \beta-\varepsilon$ for $n \geq n(\varepsilon)$. If $\beta = \infty$, then $\beta-\varepsilon$ devotes any real number $\zeta > 0$. Hence $H^*(\beta-\varepsilon, \mathcal{P}(\Omega);M_n) = \infty$

Recall that $H^*(\beta-\varepsilon,\mathcal{P}(\Omega);M_n)$ is the limit of the increasing family of "approximating" measures $M_2(h_2(\beta-\varepsilon,\cdot), \mathcal{P}(\Omega)_\alpha ;M_n) =$

$$= \inf\{ \sum_{k=1}^{\infty} (\mathrm{diam}_2 C_k)^{\beta-\varepsilon} \mid M_n \subseteq \bigcup_{k=1}^{\infty} C_k , \mathrm{diam}_2 C_k \leq \alpha \}$$

Therefore we may choose $\alpha > 0$ such that $M_2^*(h_2(\beta-\varepsilon,\cdot), \mathcal{P}_\alpha(\Omega);M_n) \geq g^{\beta-\varepsilon}$

As a covering class for $\dim_1$ we take the set $\mathcal{T}_1$ of all open spheres (with respect to the semimetric d_1). This is no restriction, according to theorem 13.6. Let $(S_k)_{k\in\mathbb{N}}$ be a covering of M_n by open spheres $S_k \in \mathcal{T}_1$, $\mathrm{diam}_1 S_k \leq \omega < \frac{1}{n}$. If no covering of this kind exists, then $\dim_1 M = \infty$ and nothing is to prove. Clearly we may assume $S_k \cap M_n \neq \emptyset$ and $S_k = K_1(x,r)$ with $x \in M_n$ (otherwise take $\frac{1}{2n}$ instead of $\frac{1}{n}$) (since deletion of other spheres only reduces the value of

$\sum_{k=1}^{\infty}\left[\operatorname{diam}_1 S_k\right]^{(\delta-\varepsilon)(\beta-\varepsilon)}$). Therefore $\left[\operatorname{diam}_1 S_k\right]^{\delta-\varepsilon} \geq d_2\ (g,S_k)$ and

$\sum_{k=1}^{\infty}\left|\operatorname{diam}_1 S_k\right|^{(\delta-\varepsilon)(\beta-\varepsilon)} \geq \sum_{k=1}^{\infty} d_2(g,S_k)^{\beta-\varepsilon}$.

Since

$d_2(g,S_k) \geq \frac{1}{g}\operatorname{diam}_2 S_k$ we obtain $M_1^*\ (h_1((\delta-\varepsilon)(\beta-\varepsilon),\cdot))\,\mathcal{T}_{1,\omega};M_n) \geq$

$\geq g^{-\beta+\varepsilon}M_2^*(h_2(\beta-\varepsilon,\cdot);\ \mathcal{P}(\Omega)_\alpha;\ M_n) \geq 1$ with $\alpha = g\omega^{\delta-\varepsilon}$ and ω small enough, such that $M_2^*(h_2(\beta-\varepsilon,\cdot),\ \mathcal{P}(\Omega)_\alpha;\ M_n) \geq g^{\beta-\varepsilon}$.

Therefore $\dim_1 M_n \geq (\delta-\varepsilon)(\beta-\varepsilon)$. Since $\varepsilon > 0$ was arbitrary $\dim_1 M_n \geq \delta\ \beta$ which gives the result.

Theorem 16.3(Billingsley [1]): Let P_1 and P_2 be two probability measures on $(\Omega,\mathcal{F})$.A stochastic process $(f_n)_{n\in\mathbb{N}}$ defines a set of cylinders. If

$\liminf\limits_{k\to\infty} \frac{\log P_2(Z(s_1,\ldots,s_k))}{\log P_1(Z(s_1,\ldots,s_k))} \geq \delta$ for all cylinders $Z(s_1,\ldots,s_k)$ which intersect M, then $\operatorname{bim}_1 M \geq \delta\ \operatorname{bim}_2 M$.

Proof: If $\operatorname{diam}_1^*(K_1(x,r)) > 0$, then we have shown in the proof of theorem 15.3 that there is a cylinder $Z(s_1,\ldots,s_k)$ with $K_1(x,r) \subseteq Z(s_1,\ldots,s_k)$ but $K_1(x,r) \not\subseteq Z(s_1,\ldots,s_k,s_{k+1})$. Hence $\operatorname{diam}_i^*\ (K_1(x,r)) = P_i(K_1(x,r))$ for $i = 1$ and 2 . Now suppose that $\operatorname{diam}_1^*\ (K_1(x,r)) = 0$. Looking again to the sequence of cylinders $Z(s_1,\ldots,s_k)$ which contain x , we find a cylinder $Z(s_1,\ldots,s_t)$ with $P_1(Z(s_1,\ldots,s_t)) = 0$, but $P_1(Z(s_1,\ldots,s_{t-1})) > 0$ (here we used again $\lim\limits_{k\to\infty} P_1(Z(s_1,\ldots,s_k) = 0$. If $P_1(Z(s_1,\ldots,s_k)) < r$, then $Z(s_1,\ldots,s_k) \subseteq K_1(x,r)$. Clearly $Z(s_1,\ldots,s_t) \subseteq K_1(x,r)$. On the other hand,if $y \in K_1(x,r)$.Therefore $d_1^*(x,y) = 0$ and $y \in Z(s_1,\ldots,s_t)$. Hence $K_1(x,r) = Z(s_1,\ldots,s_t)$ and again $\operatorname{diam}_i^* K_1(x,r) = P_i(K_1(x,r))$.

Theorem 16.4: Take $\Omega = K$, the n-dimensional unit cube, $P = \lambda$ and d the Euclidean metric. Then for any stochastic process $\dim(\mathfrak{Z};M) \leq n \operatorname{bim} M$ for any set M.

Proof: Since $\lambda(A) \leq c_n(\operatorname{diam} A)^n$ for any measurable set A with a con -

stant c_n (one can show:

$$\frac{(\frac{1}{2})^n}{2^n\ \Gamma(\frac{n}{2}+1)} = \frac{1}{\kappa_n} = c_n$$

see § 13), we have $\log \lambda(A) \le \log c_n + n \log \operatorname{diam} A$

Therefore

$$\liminf_{k\to\infty} \frac{\log \operatorname{diam} Z(s_1,\dots,s_k)}{\log \lambda(Z(s_1,\dots,s_k))} \ge \frac{1}{n} \quad \text{for any cylinder } Z(s_1,\dots,s_k) .$$

The conclusion follows in a similar manner as in theorem 16.3.

Remark: Clearly, dim $M \le n$ bim M follows immediately.

Theorem 16.5 (Wegmann [1]): Let $(f_n)_{n\in \mathbb{N}}$ be a discrete stochastic process defined on $([0,1], \mathcal{L}, \lambda)$ subject to the following conditions:

(A) $\lim_{k\to\infty} \lambda(Z(s_1,\dots,s_k)) = 0$

(B) $\lambda(Z(s_1,\dots,s_k)) > 0$

(C) Every cylinder $Z(s_1,\dots,s_k)$ is an intervall

(D) $$\lim_{k\to\infty} \frac{\log\lambda(Z(s_1,\dots,s_k))}{\log\lambda(Z(s_1,\dots,s_k,s_{k+1}))} = 1$$

for every cylinder $Z(s_1,\dots,s_{k+1})$ which intersects M,

then $\dim(\mathfrak{z};M) = \dim M$.

Proof: We denote the Euclidean metric d_E with d_1 and the semimetric induced by the process with d_2. Since $d_1(x,y) \le d_2(x,y)$ clearly $\dim_1 M = \dim M \le \dim_2 M = \dim(\mathfrak{z};M)$ (note that for the semimetric d_2 theorem 15.3 holds).

For $\varepsilon > 0$ and $k \in \mathbb{N}$ we put

$$M_k = \{x\in M \mid \frac{\log \lambda(Z(s_1,\dots,s_n))}{\log \lambda(Z(s_1,\dots,s_n,s_{n+1}))} \ge 1-\varepsilon \text{ for all } n \ge k, x\in Z(s_1,\dots,s_n)\}$$

We want to show $\dim_1 M_k \ge (1-\varepsilon)\dim_2 M_k$. This would show

$\dim_1 M \ge (1-\varepsilon)\dim_2 M$

Since $\varepsilon > 0$ can be chosen arbitrary we obtain the desired result. To this aim we will show

$$\liminf_{r\to 0} \frac{\log d_2\ (2,K_1(x,r))}{\log \operatorname{diam}_1 K_1(x,r)} \ge 1-\varepsilon \quad \text{for } x\in M_k \text{ and apply theorem 16.2.}$$

We first remark the following: If $A \in M_k$ is any set with $\mathrm{diam}_2\, A > 0$, then there are $x,y \in A$ and a cylinder $Z(s_1,\dots,s_t)$ such that

$A \subseteq Z(s_1,\dots,s_t)$

$$\lambda(Z(s_1,\dots,s_t)) = \mathrm{diam}_2\, A = d_2(x,y)$$

This follows easily from condition (A): There are only finitely many cylinders $Z(s_1,\dots,s_t)$ which contain A. Choose $Z(s_1,\dots,s_t)$ the smallest one. Then $\mathrm{diam}_2\, A \leq \lambda(Z(s_1,\dots,s_t))$.

Take $x \in A$ and a cylinder $Z(s_1,\dots,s_t,s_{t+1})$ which contains x, then there is at least one $y \in A$ such that $y \notin Z(s_1,\dots,s_t,s_{t+1})$.

Then $\quad d_2(x,y) = \lambda(Z(s_1,\dots,s_t)) \geq \mathrm{diam}_2\, A.$

Let $x \in M_k$ and $r > 0$. Then $K_1(x,r) = \,]x-r,x+r[\, \cap M_k$ (note that by theorem 13.7 we can restrict our considerations to the subspace (M_k, d_i), $i = 1,2$ in its induced semimetrics) is an open sphere in (M_k, d_1).

If $\mathrm{diam}_2\, K_1(x,r) = 0$, then $\mathrm{diam}_1\, K(x,r) = 0$ and

$$\frac{\log d_2\ (2,K_1(x,r))}{\log \mathrm{diam}_1\ K_1(x,r)} = 1$$

by our convention. If $\mathrm{diam}_2\, K_1(x,r) > 0$, then $\mathrm{diam}_1 K(x,r) > 0$ by our remark. We put $a = \inf K_1(x,r)$ and $b = \sup K_1(x,r)$ and note $\mathrm{diam}_1\, K_1(x,r) = b - a \leq 2\, r$ and $K_1(x,r) \subseteq [a, b]$

(a) If there is a cylinder $Z(s_1,\dots,s_{t+1})$ such that $Z(s_1,\dots,s_t,s_{t+1}) \subseteq [a, b]$ and $K_1(x,r) \subseteq Z(s_1,\dots,s_t)$ then $d_2(2, K_1(x,r)) \leq \mathrm{diam}_2\, K_1(x,r) \leq \lambda(Z(s_1,\dots,s_t))$

$$\mathrm{diam}_1\, K_1(x,r) \geq \lambda(Z(s_1,\dots,s_{t+1}))$$

Hence

$$\frac{\log d_2(2, K_1(x,r))}{\log \mathrm{diam}_1\, K_1(x,r)} \geq \frac{\log \lambda(Z(s_1,\dots,s_t))}{\log \lambda(Z(s_1,\dots,s_t,s_{t+1}))}$$

We may assume that x is an interior point of al cylinders containing x. This excludes an at most countable set from M which has dimension zero (for coverings by cylinders this can be seen easily from (A). Hence, if r is small enough $t \geq k$ (clearly r depends on x) and

$$\frac{\log d_2(2,\ K_1(x,r))}{\log \operatorname{diam}_1\ K_1(x,r)} \geq 1 - \varepsilon$$

(b) We now suppose that there is no cylinder $Z(s_1,\ldots,s_{t+1})$ such that $Z(s_1,\ldots,s_{t+1}) \subseteq [a,\ b]$ and $K_1(x,r) \subseteq Z(s_1,\ldots,s_t)$

We now define

$A_1 = \{y \in K_1(x,r) \mid$ There is a cylinder $Z(s_1,\ldots,s_m)$ such that $y \in Z(s_1,\ldots,s_m)$, $a \in Z(s_1,\ldots,s_m)$, $b \notin Z(s_1,\ldots,s_m)\}$

$A_2 = \{z \in K_1(x,r) \mid$ There is a cylinder $Z(s_1',\ldots,s_n')$ such that $z \in Z(s_1',\ldots,s_n')$, $a \notin Z(s_1',\ldots,s_n')$, $b \in Z(s_1',\ldots,s_n')\}$

Since $n \leq m$ implies $Z(s_1,\ldots,s_m) \subseteq Z(s_1',\ldots,s_n')$ or $Z(s_1,\ldots,s_m) \cap Z(s_1',\ldots,s_n') = \emptyset$, we see $A_1 \cap A_2 = \emptyset$. Since for every $t \in K_1(x,r)$ the sequence of cylinders $Z(s_1,\ldots,s_m)$ which contain t is a contracting sequence of intervalls, we see $A_1 \cup A_2 = K_1(x,r)$.

(b')Now suppose $\operatorname{diam}_2 A_1 > 0$. Therefore we can find $y,u \in A_1$ and a cylinder $Z(s_1,\ldots,s_m)$ such that $\operatorname{diam}_2 A_1 = d_2(y,u) = \lambda(Z(s_1,\ldots,s_m))$ and $A_1 \subseteq Z(s_1,\ldots,s_m)$, $u < y$. If $y \notin Z(s_1,\ldots,s_m,s_{m+1})$ then (by conditions (A) and (B)) $\lambda Z(s_1,\ldots,s_m,s_{m+1}))<\lambda(Z(s_1,\ldots,s_m))=d_2(y,u)$ and therefore $u \in Z(s_1,\ldots,s_m,s_{m+1})$ and (since $a \leq u$ and the cylinders are intervalls) $a \notin Z(s_1,\ldots,s_m,s_{m+1})$.

The very property $y \in A_1$ ensures $b \notin Z(s_1,\ldots,s_m,s_{m+1})$ and therefore $Z(s_1,\ldots,s_m,s_{m+1}) \subseteq [a,\ b]$.

If $\operatorname{diam}_2 A_2 = 0$, then again

$$d_2(2,\ K_1(x,r)) \leq \operatorname{diam}_2\ A_1 \leq \lambda(Z(s_1,\ldots,s_m))$$

and $\operatorname{diam} K_1(x,r) \geq \lambda(Z(s_1,\ldots,s_m,s_{m+1}))$ and we conclude as in the foregoing case (a) .

(b'') If additionally $\operatorname{diam}_2 A_2 > 0$ we can find similarly a cylinder $Z(s_1',\ldots,s_n',s_{n+1}')$ such that $A_2 \subseteq Z(s_1',\ldots,s_n')$ and

$Z(s_1',\dots,s_n',s'_{n+1}) \subsetneq [a,\ b]$

Clearly $Z(s_1',\dots,s_{n+1}') \cap Z(s_1,\dots,s_{m+1}) = \emptyset$.

Hence

$$d_2(2,\ K_1(x,r)) \le \operatorname{diam}_2 A_1 + \operatorname{diam}_2 A_2 \le \lambda(Z(s_1,\dots,s_m)) + \lambda(Z(s_1',\dots,s_n'))$$

$\operatorname{diam}_1 K_1(x,r) \ge \lambda(Z(s_1,\dots,s_{m+1})) + \lambda(Z(s_1',\dots,s_{n+1}'))$ and

$$\frac{\log d_2(2,K_1(x,r))}{\log \operatorname{diam}_1 K_1(x,r)} \ge \frac{\log\left[\lambda(Z(s_1,\dots,s_m)) + \lambda(Z(s_1',\dots,s_n'))\right]}{\log\left[\lambda(Z(s_1,\dots,s_{m+1})) + \lambda(Z(s_1',\dots,s_{n+1}'))\right]}$$

Choosing r sufficiently small, we can again assume $m,n \ge k$. Therefore

$$\frac{\log \lambda(Z(s_1,\dots,s_m))}{\log \lambda(Z(s_1,\dots,s_{m+1}))} \ge 1-\varepsilon$$

and

$$\frac{\log \lambda(Z(s_1',\dots,s_n'))}{\log \lambda(Z(s_1',\dots,s_{n+1}'))} \ge 1-\varepsilon$$

which both imply

$$\frac{\log d_2(2,K_1(x,r))}{\log \operatorname{diam}_1 K_1(x,r)} \ge 1-\varepsilon + \frac{\log 2}{\log\left[\lambda(Z(s_1,\dots,s_{m+1})) + \lambda(Z(s_1',\dots s_{n+1}'))\right]}$$

Since $\lim_{k\to\infty} \lambda(Z(t_1,\dots,t_k)) = 0$ we obtain

$$\liminf_{r\to 0} \frac{\log d_2(2,K_1(x,r))}{\log \operatorname{diam}_1 K_1(x,r)} \ge 1-\varepsilon .$$

(b''') The case $\operatorname{diam}_2 A_1 = 0$ and $\operatorname{diam}_2 A_2 > 0$ is a simple duplication of (b'). If, at last, $\operatorname{diam}_2 A_1 = \operatorname{diam}_2 A_2 = 0$, then $\operatorname{diam}_2 K_1(x,r)=0$, which case was investigated at the very beginning.

§ 17. The main theorem of dimension theory of Jacobi algorithm

Here we return to Jacobi algorithm. The following theorem traces back to the pioneer work of Jarnik [1] and Good [1] and was worked out in Schweiger [8] and Schweiger-Stradner [1], [2]. In some aspects similar

results can be found in Šalát [1], Jakubec [3] and Smorodinsky [1].
Let $R \subsetneq \mathbb{N}_0^n$, $R \neq \emptyset$, (recall $\mathbb{N}_0 = \mathbb{N} \cup \{0\}$) and define

$$E(R) = \bigcap_{s=1}^{\infty} \bigcup_{k_1,..,k_s \in R} B(k_1,\ldots,k_s)$$

If the sequence $(k_1,\ldots,k_s)$ is not admissible, we put $B(k_1,\ldots,k_s) = \emptyset$.
If $R = I = \{a = (a_1,\ldots,a_n) \mid 0 \leq a_i \leq a_n \ , \ 1 \leq a_n\}$,
then clearly

$$E(I) = B$$

When R is a proper subset of I we can conclude

$$\lambda(E(R)) = 0$$

Since λ - almost every $x \in B$ contains every digit in its development (and by the pointwise ergodic theorem the frequency of a digit k is $\mu(B(k))$). We now impose the following restriction on the set R:

(P) Every cylinder $B(k_1,\ldots,k_q)$ where the digits $k_1,\ldots,k_q$ are taken from R is proper (that means $T^q B(k_1,\ldots,k_q) = B$).

Then we can formulate the main

<u>Theorem 17.1:</u> If a set $R \subseteq I$ satisfies condition (P), then we can calculate bim E(R) as follows: Let

$$x_q = \inf \{x \in [0,1] \mid \sum_{k_1,..,k_q \in R} \lambda(B(k_1,\ldots,k_q))^x \leq 1\}$$

then $\lim_{q\to\infty} x_q$ exists and

$$\text{bim } E(R) = \lim_{q\to\infty} x_q$$

<u>Remark 1:</u> Since $\sum_{k_1,..,k_q \in R} \lambda(B(k_1,\ldots,k_q)) \leq 1$, the set of all $x \in [0,1]$, such that $\sum_{k_i \in R} \lambda(B(k_1,\ldots,k_q))^x \leq 1$ is not empty and $x_q \leq 1$.

<u>Remark 2:</u> The theorem seems to be true for all subsets $R \subseteq I$, but no proof is available.

<u>Proof</u>: The proof of this theorem is somewhat lengthy and will be divide into several steps.
We introduce the three series

$$f_q(x) = \sum_{k_1,..,k_q \in R} \lambda(B(k_1,\ldots,k_q))^x$$

$$\phi_q(x) = \sum_{k_1,\dots,k_q \in R} (\sup \Delta(k_1,\dots,k_q))^x$$

$$\Phi_q(x) = \sum_{k_1,\dots,k_q \in R} (\inf \Delta(k_1,\dots,k_q))^x$$

Proposition 17.2: There is a real number $d \in [0,1]$ not dependent on q such that the series $f_q(x)$, $\phi_q(x)$ and $\Phi_q(x)$ are convergent on $]d,1]$ and divergent on $[0,d[$ (which set may be possibly empty).

Remark 1: It is not unexpected that we can say nothing about convergence or divergence for $x = d$. Clearly, when R is finite, $d = 0$ and $f_q(0) \geq 1$ is finite.

Remark 2: Clearly, $x_q \geq d$ for any $q \geq 1$.

Proof: From lemma 2.6 we know

$$c_1 \sup \Delta(k_1,\dots,k_q) \leq \lambda(B(k_1,\dots,k_q)) \leq c_2 . \inf \Delta(k_1,\dots,k_q)$$

with constants c_1, $c_2 > 0$.

Therefore the convergence of one of the three series implies the convergence of both the others. Since $x < y$ implies $\lambda(B(k_1,\dots,k_q))^y < \lambda(B(k_1,\dots,k_q))^x$ we can conclude that convergence of $f_q(x)$ implies convergence of $f_q(y)$. We put $d(q) = \inf \{x \mid f_q(x)$ is convergent$\}$

Corollary 2.8 shows

$$c_3 \lambda(B(k_1 \dots,k_q))\lambda(B(k_{q+1})) \leq \lambda(B(k_1,\dots,k_q,k_{q+1})) \leq$$

$$\leq c_4 \lambda(B(k_1,\dots,k_q))\lambda(B(k_{q+1}))$$

which gives

$$c_3^x f_q(x) f_1(x) \leq f_{q+1}(x) \leq c_4^x f_q(x) f_1(x)$$

From this relation we see by induction $d(q) = d(q+1)$ for all $q \geq 1$, hence $d(q) = d$.

Corollary 17.3: If $d < x_q$ then $f_q(x_q) = 1$ and x_q is the unique solution of $f_q(x) = 1$.

Proof: Under this assumption the series $f_q(x)$ is uniformly convergent on $[x_q,1]$ and defines a continuous, strictly decreasing function.

We now define

$$y_q = \inf \{x \in [0,1] \mid \phi_q(x) \leq 1\}$$

Recall that

$$x_q = \inf \{x \in [0,1] \mid f_q(x) \leq 1\}$$

We first prove:

Proposition 17.4: If $\lim_{q\to\infty} y_q$ exists, then $\lim_{q\to\infty} x_q$ exists also and conversely.

Furthermore $\lim_{q\to\infty} y_q = \lim_{q\to\infty} x_q$

Proof: We note

$$\lambda(B(k_1,\ldots,k_q)) \leq \sup \Delta(k_1,\ldots,k_q)) \leq c_5\lambda(B(k_1,\ldots,k_q))$$

Therefore $\Phi_q(x) \leq 1$ implies $f_q(x) \leq 1$. Hence $x_q \leq y_q$. Given $\varepsilon > 0$ we see that (note that $c_5^{\,t} \leq c_5$ for $0 \leq t \leq 1$):

$$\Phi_q(x_q + 2\varepsilon) \leq c_5\, f_q(x_q + 2\varepsilon) \leq c_5\, f_q(x_q + \varepsilon)\ .$$

$$.\ [\max \lambda(B(k_1,\ldots,k_q))^{\varepsilon} < c_5[\max\ \lambda(B(k_1,\ldots,k_q))]^{\varepsilon} < 1$$

when $q \geq q(\varepsilon)$. This shows $y_q \leq x_q + 2\varepsilon$ for $q \geq g(\varepsilon)$.

Defining

$$z_q = \inf \{x \in [0,1] \mid \Phi_q(x) \leq 1\}$$

and using

$$\lambda(B(k_1,\ldots,k_q)) \geq c_6 \inf \Delta(k_1,\ldots,k_q) \geq c_7\lambda(B(k_1,\ldots,k_q))$$

a similar result holds for $(z_q)_{q\in\mathbb{N}}$. Since $c_6 < 1$ the proof parallels the second part of the proof of proposition 17.4.

Def.: An R-admissible sequence of digits will be an admissible sequence the digits of which are all taken from R.

Proposition 17.5: $\lim_{q\to\infty} y_q$ exists

Proof: Since

$$\sup \Delta(k_1,\ldots,k_q,a_1,\ldots,a_p) \leq \sup \Delta(k_1,\ldots,k_q)\sup \Delta(a_1,\ldots,a_p)$$

we obtain by summing left over all R-admissible sequences $k_1,\ldots,k_q$, $a_1,\ldots,a_p$ and by summing right over all pairs of R-admissible sequences $k_1,\ldots,k_q$ and $a_1,\ldots,a_p$

$$\Phi_{q+p}(x) \leq \Phi_q(x)\ \Phi_p(x)$$

Taking $x > \max(y_q,y_p)$ we obtain

$$\Phi_{q+p}(x) < 1$$

Hence $y_{p+q} \leq \max(y_q, y_p)$. This shows $y_n \leq y_1$ for all $n \geq 1$. By induction one obtains additionally

$$y_{qh} \leq y_q \quad \text{for } h \geq 1$$

and hence

$$y_{qh+pk} \leq \max(y_q, y_p)$$

We put $A = \limsup_{n \to \infty} y_n$. Note that $y_1 \geq A$. Suppose there are two different prime numbers p and q, such that $y_p < A$, $y_q < A$. Since any $n \geq n(p,q)$ is representable in the form $n = qh+pk$ with $h,k \geq 1$, we would reach

$$y_n \leq \max(y_q, y_p) < A$$

for all $n \geq n(p,q)$. This would be a contradiction. Therefore $y_p \geq A$ for all prime numbers p with the possible exception of a single prime p^*. As the example $a_k = 0$ if $k \equiv 0 \pmod 2$, $a_k = 1$ if $k \equiv 1 \pmod 2$ shows the condition $a_{p+q} \leq \max(a_p, a_q)$ does not ensure convergence of a sequence $(a_k)_{k \in \mathbb{N}}$. It is conjectured that in our special case the sequence $(y_k)_{k \in \mathbb{N}}$ is decreasing (and then convergence would follow immediately), but no proof is known. Therefore we need some extra considerations. The proof presented here is simpler than the device of Good [1] (which was also used in Schweiger [8]). It is surprising that in this connection we need the restriction (P) on the digits.

We now use

$$\sup \Delta(k_1,\dots,k_q,a_1,\dots,a_p) \geq c_8 \sup \Delta(k_1,\dots,k_q) \sup \Delta(a_1,\dots,a_p)$$

Now condition (P) ensures that the sequence $k_1,\dots,k_q, a_1,\dots,a_p$ is R-admissible when and only when $k_1,\dots,k_q$ and $a_1,\dots,a_p$ are both R-admissible sequences (if condition (P) does not hold, $k_1,\dots,k_q$ and $a_1,\dots,a_p$ may be R-admissible without being $k_1,\dots,k_q,a_1,\dots,a_p$ R-admissible). Now summing up we obtain $\Phi_{p+q}(x) \geq c_8\, \Phi_p(x)\, \Phi_q(x)$.

We put $x = y_{p+q} + 2\varepsilon$ and proceed as in proposition 17.4:

$$\Phi_q(y_{p+q}\ 2\varepsilon)\Phi_p(y_{p+q} + 2\varepsilon) \leq \frac{1}{c_8}\, \Phi_{p+q}(y_{p+q} + \varepsilon\ )\left[\max\lambda(B(k_1,\dots,a_{p+q}))\right]^{\varepsilon} < 1$$

if $p+q \geq n(\varepsilon)$. Therefore $\min(y_p, y_q) \leq y_{p+q} + 2\varepsilon$, $p+q \geq n(\varepsilon)$

Now let $y_n < A$. If $y_{n-1} < A$, we would reach (using the fact that n and n-1 are clearly relative prime) a contradiction in a similar manner as before. Therefore

$$y_n \geq \min(y_{n-1}, y_1) - 2\varepsilon \geq A - 2\varepsilon$$

for any $n \geq n(\varepsilon)$. This shows $A = \lim_{n\to\infty} y_n$

Additionally we can prove $y_n \geq A$

If $y_n < A$, we put $A - y_n = \varepsilon > 0$ and choose a prime such that $y_p \geq A$ and $y_p - y_{np} < \varepsilon$. Then

$$y_p \geq A = y_n + \varepsilon \geq y_{np} + \varepsilon$$

is a contradiction. Similarly we could show $\lim_{n\to\infty} z_n = A$ and $z_n \leq A$ for $n \leq 1$.

<u>Proposition 17.6</u>: bim $E(R) \leq A$

<u>Proof</u>: Choose any $\gamma > A$. Then we have $x_q < \gamma$ where $q \geq q(\gamma)$. Then $f_q(\gamma) < 1$. This means

$$\sum_{k_1,\ldots,k_q \in R} \lambda(B(k_1,\ldots,k_q))^\gamma < 1$$

Since $\lambda(B(k_1,\ldots,k_q)) \leq \alpha$ for $q > q(\alpha)$, we have (using the semimetric d_j^*):

$$M^*(h(\gamma,\cdot), \mathfrak{Z}_\alpha; E(R)) < 1$$

and therefore

$$H^*(\gamma, \mathfrak{Z}; E(R)) \leq 1$$

This implies bim $E(R) \leq \gamma$. Since $\gamma > A$ was arbitrary, we see bim $E(R) \leq A$.

<u>Proposition 17.7</u>: If R is a finite set, then bim $E(R) \geq A$.

<u>Note</u>: Together with proposition 17.6 we can conclude bim $E(R) = A$ if R is finite.

<u>Proof</u>: Since $A = \lim_{q\to\infty} x_q$ we can conclude

$$\sum_{k_1,\ldots,k_q \in R} \lambda(B(k_1,\ldots,k_q))^{A-\varepsilon} \geq 1$$

for any $q \geq q(\varepsilon)$. If $A = 0$ then clearly bim $E(R) \geq A$ and nothing is to be shown. Now let $Y(N,R)$ be a covering of $E(R)$ by cylinders $B(k_1,\ldots,k_s)$, where $s \geq N$ and $k_j \in R$, $1 \leq j \leq s$. Clearly we can restrict our attention to such coverings, because:

(a) If $B(k_1,\ldots,k_s) \cap E(R) = \emptyset$ we may drop this cylinder from the covering and will at least reduce the approximating measure. We can furthermore assume that the cylinders of the covering are pairwise disjoint.

(b) Since R is finite, $\min_{k_1,..,k_s \in R} \lambda(B(k_1,\dots,k_s)) > 0$ exists for fixed s. Hence by adjusting $\alpha > 0$ appropriate all cylinders of an α-covering $\mathcal{Z}_\alpha$ must have orders at least $N = N(\alpha)$.

By the estimates frequently used before we see

$$\sum_{k_1,..,k_q \in R} \lambda(B(a_1,\dots,a_r,k_1,\dots,k_q))^{A-\varepsilon} \geq$$

$$\geq c_9\ \lambda(B(a_1,\dots,a_r))^{A-\varepsilon} \sum_{k_1,..,k_q \in R} \lambda(B(k_1,\dots,k_q))^{A-\varepsilon} \geq$$

$$\geq c_9\ \lambda(B(a_1,\dots,a_r))^{A-\varepsilon}$$

On the other hand

$$\lambda(B(a_1,\dots,a_r,k_1,\dots,k_q)) \leq c_{10}\lambda(B(a_1,\dots,a_r))\tau(q)$$

where $\tau(q) = \max \lambda(B(k_1,\dots,k_q))$ tends to zero as q increases to infinity. Therefore

$$\sum_{k_1,..,k_q \in R} \lambda(B(a_1,\dots,a_r,k_1,\dots,k_q))^{A-2\varepsilon} \geq$$

$$\geq \sum_{k_1,..,k_q \in R} \lambda(B(a_1,\dots,a_r,k_1,\dots,k_q))^{A-\varepsilon}\lambda(B(a_1,\dots,a_r,k_1,\dots,k_q))^{-\varepsilon} \geq$$

$$\geq c_9\ \lambda(B(a_1,\dots,a_r))^{A-\varepsilon}\ c_{10}{}^{-\varepsilon}\ \lambda(B(a_1,\dots,a_r))^{-\varepsilon}.\tau(q)^{-\varepsilon} \geq$$

$$\geq c_{11}\ \tau(q)^{-\varepsilon}\ \lambda(B(a_1,\dots,a_r))^{A-2\varepsilon} \geq \lambda(B(a_1,\dots,a_r))^{A-2\varepsilon}$$

if $q \geq q(\varepsilon)$ since $\tau(q) \to 0$ as $q \to \infty$. We first study a covering $Y(N,R)$ of $E(R)$ satisfying (a) and (b) such that the orders s are multiples of $N \geq q(\varepsilon)$. Let us denote with $\mathcal{Z}(N,R)$ the set of all cylinders of order N the digits are taken from R.

We want to show

$$\sum_{Z \in Y(N,R)} \lambda(Z)^{A-2\varepsilon} \geq 1$$

We now construct a sequence of coverings as follows: $Y^2(N,R)$ consists of all cylinders from $Y(N,R) \cap \mathfrak{Z}(N,R)$ and the set of cylinders $\mathfrak{Z}'(2N,R) \subseteq \mathfrak{Z}(2N,R)$ which are necessary to cover $E(R)$. Then

$$\sum_{z \in Y^2(N,R)} \lambda(z)^{A-2\varepsilon} = \sum_{z \in \mathfrak{Z}(N,R) \cap Y(N,R)} \lambda(z)^{A-\varepsilon} + \sum_{z \in \mathfrak{Z}'(2N,R)} \lambda(z)^{A-2\varepsilon}$$

Since all cylinders of $Y^2(N,R)$ are disjoint and their union covers $E(R)$ we see

$$\sum_{z \in \mathfrak{Z}'(2N,R)} \lambda(z)^{A-2\varepsilon} \geq \sum_{z \in \mathfrak{Z}(N,R) \setminus Y(N,R)} \lambda(z)^{A-2\varepsilon}$$

Hence

$$\sum_{z \in Y^2(N,R)} \lambda(z)^{A-2\varepsilon} \geq \sum_{z \in \mathfrak{Z}(N,R)} \lambda(z)^{A-2\varepsilon} \geq 1$$

$Y^3(N,R)$ consists of all cylinders from $Y(N,R) \cap \mathfrak{Z}(N,R)$ and $Y(N,R) \cap \mathfrak{Z}(2N,R)$ and the set of cylinders $\mathfrak{Z}'(3N,R) \subset \mathfrak{Z}(3N,R)$ necessary to give a covering. Then

$$\sum_{z \in Y^3(N,R)} \lambda(z)^{A-2\varepsilon} = \sum_{z \in \mathfrak{Z}(N,R) \cap Y(N,R)} \lambda(z)^{A-2\varepsilon} + \sum_{z \in \mathfrak{Z}(2N,R) \cap Y(N,R)} \lambda(z)^{A-2\varepsilon} +$$

$$+ \sum_{z \in \mathfrak{Z}'(3N,R)} \lambda(z)^{A-2\varepsilon} \geq \sum_{z \in \mathfrak{Z}(N,R) \cap Y(N,R)} \lambda(z)^{A-2\varepsilon} + \sum_{z \in \mathfrak{Z}'(2N,R)} \lambda(z)^{A-2\varepsilon} \geq 1$$

By induction we obtain a covering $Y^h(N,R)$ consisting of all cylinders from $Y(N,R)$ the order of which is $\leq h.N$ and the set $\mathfrak{Z}'(h.N,R) \subseteq \mathfrak{Z}((h.N,R)$ which fills up to a covering. Then

$$\sum_{z \in Y^h(N,R)} \lambda(z)^{A-2\varepsilon} \geq 1$$

We now use the following result the proof of which will be postponed (see proposition 17.8): If R is a finite set, then any covering by disjoint cylinders the digits of which are taken from R, is finite.

By proposition 17.8 $Y(N,R)$ is finite and therefore $Y(N,R) = Y^h(N,R)$ for $h \geq 1$.

Now let $X(N,R)$ be an arbitrary covering, where $N \geq q_1(\varepsilon) > q(\varepsilon)$. Let $B(k_1,\ldots,k_s) \in X(N,R)$. If $s \equiv 0 \pmod{q(\varepsilon)}$ we retain the cylinder. If not we substitute $B(k_1,\ldots,k_s)$ by the set of cylinders $B(k_1,\ldots,k_s,a_1,\ldots,a_r)$ where $a_i \in R$, $1 \leq i \leq r$ and $s+r \equiv 0 (\mathrm{mod}\ q(\varepsilon))$, but $r < q(\varepsilon)$. We see

$$\sum_{a_1,\dots,a_r \in R} \lambda(B(k_1,\dots,k_s,a_1,\dots,a_r))^{A-2\varepsilon} \le$$

$$\le c_{11}\tau(s)^{+\varepsilon}\lambda(B(k_1,\dots,k_s))^{A-3\varepsilon} \cdot \sum_{a_1,\dots,a_r \in R} \lambda(B(a_1,\dots,a_r))^{A-3\varepsilon} \le$$

$$\le \lambda(B(k_1,\dots,k_s))^{A-3\varepsilon}$$

if $s \ge N \ge q_1(\varepsilon)$. Hence we have

$$\sum_{z \in X} \lambda(z)^{A-3\varepsilon} \ge \sum_{z \in X'} \lambda(z)^{A-2\varepsilon} \ge 1$$

where X' is the modified covering. Therefore we result in

$$\sum_{z \quad X} \lambda(z)^{A-3\varepsilon} \ge 1$$

for any covering X the lowest order of which is $\ge q_1(\varepsilon)$.

Taking $q_1(\varepsilon) \ge N(\alpha)$, we obtain

$$\operatorname{bim} E(R) \ge A - 3\varepsilon$$

Since $\varepsilon > 0$ was arbitrary we are done.

Proposition 17.8: If R is finite, every covering by pairwise disjoint cylinders the digits of which are taken from R is finite.

Proof: The proof given is an adaption of the well-known Tychonoff theorem on the product of compact spaces. We introduce the sets

$$Z(s,k) = \{x \mid k_s(x) = k\}$$

Clearly

$$B(a_1,\dots,a_s) = Z(1,a_1) \cap \dots\dots \cap Z(s,a_a)$$

It is enough to show that every cover of E(R) by a subfamily of $\{Z(s,k) \mid s \in \mathbb{N},\ k \in R\}$ contains a finite subcover. Then every covering by cylinders contains a finite subcover. Since a covering by disjoint cylinders cannot be refined, proposition 17.8 is shown. Now let S be a subfamily of $\{Z(s,k) \mid s \in \mathbb{N},\ k \in R\}$ which does not contain a finite subcollection covering E(R).

Then (since R is finite)

$$O(s) = \{k \in R \mid Z(s,k) \in S\} \neq R$$

Therefore we can select $k_s \notin O(s)$. By condition (P) the sequence $(k_1,k_2,\dots)$ is admissible and defines a point not covered by any set of S.

<u>Proposition 17.9</u>: Let R be infinite and $(R_m)_{m\in\mathbb{N}}$ be an increasing sequence of finite sets of digits such that $\bigcup_{m=1}^{\infty} R_m = R$, then

$$\lim_{m\to\infty} \text{bim } E(R) \leq \text{bim } E(R)$$

<u>Proof</u>: This is only a reformulation of lemma 14.4. Note that $E(R_m) \subseteq E(R_{m+1})$ implies that $(\text{bim } E(R_m))_{m\in\mathbb{N}}$ is an increasing sequence. Since $\bigcup_{m=1}^{\infty} E(R_m) \neq E(R)$ we cannot conclude equality at this point.

Now we put

$$x_q(m) = \inf \{x \in [0,1] | \sum_{k_1,\ldots,k_q)\in R_m} \lambda(B(k_1,\ldots,k_q))^x \leq 1\}$$

By propositions 17.4 and 17.5 we know that $\lim_{q\to\infty} x_q(m) = A(m)$ exists for every set R_m. By propositions 17.6 and 17.7 we can conclude that $\text{bim } E(R_m) = A(m)$. Our theorem will be proved if we can show

<u>Proposition 17.1o</u>: $\lim_{m\to\infty} A(m) = A$

<u>Proof</u>: We put

$$z_q(m) = \inf \{x \in [0,1] | \sum_{k_1,\ldots,k_q\in R_m} \inf \Delta(k_1,\ldots,k_q)^x \leq 1\}$$

The inequality

$$\Phi_{q+m}(x) \geq \Phi_q(x)\Phi_m(x)$$

shows $z_{q+m} \geq \min(z_q,z_m)$. Similar to proposition 17.5 and by the remark following proposition 17.4 we see $\lim_{q\to\infty} z_q(m) = A(m)$ and $z_q(m) \leq A(m)$ for all $q \geq 1$. From

$$\sum_{k_1,\ldots,k_q\in R_m} \inf \Delta(k_1,\ldots,k_q)^x \leq \sum_{k_1,\ldots,k_q\in R_{m+1}} \inf \Delta(k_1,\ldots,k_q)^x$$

we see that $z_q(m)$ is monotonically increasing to a limit $a_q \leq z_q$.

Then

$$1 = \sum_{k_1,\ldots,k_q\in R_m} \inf \Delta(k_1,\ldots,k_q)^{z_q(m)} \geq \sum_{k_1,\ldots,k_q\in R_m} \inf\Delta(k_1,\ldots,k_q)^{a_q}$$

Since this estimate holds for all m, we obtain $z_q \le a_q$ hence $\lim_{m\to\infty} z_q(m) = z_q$. From $z_q(m) \le z_q(m+1) \le z_q$ we see $A(m) = \lim_{q\to\infty} z_q(m) \le$ $\le A(m+1) \le A$. We put $A' = \lim_{m\to\infty} A(m)$. Hence $A' \le A$.

Now for $q \ge 1$ we see

$$\sum_{k_1,\ldots,k_q \in R_m} \inf \Delta(k_1,\ldots,k_q)^{A'} \le \sum_{k_1,\ldots,k_q \in R_m} \inf \Delta(k_1,\ldots,k_q)^{A(m)} \le$$

$$\le \sum_{k_1,\ldots,k_q \in R_m} \inf \Delta(k_1,\ldots,k_q)^{z_q(m)} = 1$$

Since this estimate holds for all m, we see

$$\sum_{k_1,\ldots,k_q \in R} \inf \Delta(k_1,\ldots,k_q)^{A'} \le 1$$

Hence $z_q \le A'$. Since we can choose q arbitrary we obtain $A \le A'$.

<u>Corollary 17.11</u>: $\lim_{m\to\infty} \text{bim } E(R_m) = \text{bim } E(R)$

<u>Proof</u>: This follows at once from proposition 17.1o. This corollary is remarkable since $\bigcup_{m=1}^{\infty} E(R_m) \subseteq E(R)$ and this inclusion is proper.

§ 18. Further results on Billingsley dimension

We first present the following generalization of proposition 17.8:

<u>Theorem 18.1:</u> Let $R \subseteq I$ be any finite set of digits, then every covering of E(R) by disjoint cylinders the digits of which are taken from R is finite.

<u>Remark</u>: The novelty is that we do not use condition (P). In § 17 we gave another proof (which was an easy translation of Tychonoff's theorem) using condition (P). Theorem 18.1 was used steadily, but tacitly assumed in Schweiger [8] .

<u>Proof</u>: Let X be any covering of E(R) by disjoint cylinders, $X = \{Z\}$. If X is not finite there is at least one $k_1 \in R$ such that $B(k_1) \cap E(R)$ is not covered by finitely many cylinders $Z \in X$. Repeating this argument there is a $k_2 \in R$ such that $B(k_1,k_2) \cap E(R)$ has the same proper-

ty. In this way we find a sequence $(k_1,k_2,\dots)$ such that $B(k_1,\dots,k_s) \cap E(R)$ is not covered by finitely many cylinders $Z \in X$. Since $B(k_1,\dots,k_s) \neq \emptyset$ for any $s \geq 1$ the sequence $(k_1,k_2,\dots)$ is admissible and defines a point $x = \Psi(k_1,k_2,\dots)$. Clearly $x \in E(R)$ (note that $E(R)$ is a perfect set) and hence there is a cylinder $Z = B(k_1,\dots,k_s)$ which contains x. Then $B(k_1,\dots,k_s) \cap E(R) \subseteq B(k_1,\dots,k_s)$ is finitely covered.

Theorem 18.2: Let $R \subseteq I$ be finite.

(a) Suppose that for σ, $0 < \sigma \leq 1$, there exists a $\nu(\sigma)$, such that

$$\sum_{k_{m+1} \in R} \lambda(B(k_1,\dots,k_m,k_{m+1}))^\sigma \geq \lambda(B(k_1,\dots,k_m))^\sigma \quad \text{for } m \geq \nu(\sigma) ,$$

then $\operatorname{bim} E(R) \geq \sigma$.

(b) Suppose that for σ, $0 < \sigma \leq 1$ there exists a $\nu(\sigma)$, such that

$$\sum_{k_{m+1} \in R} \lambda(B(k_1,\dots,k_m,k_{m+1}))^\sigma \leq \lambda(B(k_1,\dots,k_m))^\sigma \quad \text{for } m \geq \nu(\sigma) ,$$

then $\operatorname{bim} E(R) \leq \sigma$.

Remark: If $k_1,\dots,k_m,k_{m+1}$ is not admissible, we put $B(k_1,..,k_m,k_{m+1})=\emptyset$.

Proof:

(a) Similarly to proposition 17.7 we obtain

$$\sum_{Z \in X} \lambda(Z)^\sigma \geq \sum_{k_1,\dots,k_\nu \in R} \lambda(B(k_1,\dots,k_\nu))^\sigma > 0$$

for any covering, if all cylinders have order $\geq \nu = \nu(\sigma)$. Hence $\operatorname{bim} E(R) \geq \sigma$.

(b) Clearly

$$\sum_{k_1,\dots,k_s \in R} \lambda(B(k_1,\dots,k_s))^\sigma \leq \sum_{k_1,\dots,k_\nu \in R} \lambda(B(k_1,\dots,k_\nu))^\sigma < \infty$$

for any $s \geq \nu = \nu(\sigma)$. This implies $\operatorname{bim} E(R) \leq \sigma$.

Theorem 18.2 can be used to give estimates on Billingsley dimension of the set $E(R)$ (for R finite) even if the set R does not satisfy the assumption (P). One notes that condition (P) was only used in proposition 17.5 (the proof of proposition 17.8 may be replaced by the proof of theorem 18.1). Therefore it seems to be likely that theorem 17.1 holds without any further assumption. Theorem 18.2 allows to prove several generalizations of theorems found by Jarnik [1] and Good [1] (cf. Schweiger [8]).

We will list some cases and corollaries. We first recall corollary 2.8 for a special case:

$$c_1\lambda(B(a_1,..,a_s))\lambda(B(a_{s+1})) \le \lambda(B(a_1,..,a_s,a_{s+1})) \le c_2\lambda(B(a_1,..,a_s)) \cdot$$

$$\cdot\ \lambda(B(a_{s+1}))$$

<u>Theorem 18.3:</u> For any finite set $R \subseteq I$ we can find constants $c_4(R)$, $c_5(R)$ such that

$$1 - c_4(R) \sum_{a \in I\setminus R} \lambda(B(a)) \le \text{bim } E(R) \le 1 - c_5(R) \sum_{a \in I\setminus R} \lambda(B(a))$$

provided that $c_2 \sum_{a \in I\setminus R} \lambda(B(a)) < 1$ which is true for R not too small.

<u>Proof</u>: We begin with the basic relation (corollary 2.8):

$$c_1\lambda(B(k_1,..,k_q))\lambda(B(k_{q+1})) \le \lambda(B(k_1,..,k_q,k_{q+1})) \le$$

$$\le c_2\lambda(B(k_1,..,k_q))\lambda(B(k_{q+1}))$$

and

$$\lambda(B(k_1,...,k_q)) = \sum_{k_{q+1}\in I} \lambda(B(k_1,..,k_q,k_{q+1}))$$

This shows further

$$\lambda(B(k_1,...,k_q)) = \sum_{k_{q+1}\in R} \lambda(B(k_1,..,k_q,k_{q+1})) + \sum_{k_{q+1}\in I\setminus R} \lambda(B(k_1,..,k_q,k_{q+1})) \le$$

$$\le \sum_{k_{q+1}\in R} \lambda(B(k_1,..,k_q,k_{q+1})) + \lambda(B(k_1,..,k_q))c_2 \sum_{k_{q+1}\in I\setminus R} \lambda(B(k_{q+1}))$$

Here nothing is lost if we sum over all $k_{q+1}\in I\setminus R$ irrespectively if $(k_1,...,k_q,k_{q+1})$ is admissible. Hence

$$\sum_{k_{q+1}\in R} \lambda(B(k_1,..,k_q,k_{q+1})) \ge \lambda(B(k_1,..,k_q))(1 - c_2 \sum_{k_{q+1}\in I\setminus R} \lambda(B(k_{q+1}))$$

Since $\lambda(B(k_1,...,k_q,k_{q+1})) < c_3\lambda(B(k_1,...,k_q))$, $0 < c_3 < 1$, we obtain

$$\lambda(B(k_1,...,k_q,k_{q+1}))^{\sigma-1} > c_3^{\sigma-1}\ \lambda(B(k_1,...,k_q))^{\sigma-1}$$

The relation used before can be proved as follows:

First note that

$$d = \min \{ \sum_{a \neq k_{q+1}} \lambda(B(a)) \mid (k_1,..,k_q,k_{q+1}) \text{ is admissible for every}$$
k_{q+1}

choice of $(k_1,...,k_q)\}$ is an absolute constant, $0 < d < 1$. Then

$$\lambda(B(k_1,..,k_q)) = \sum_{a \neq k_{q+1}} \lambda(B(k_1,...,k_q,a)) + \lambda(B(k_1,...,k_q,k_{q+1})) \geq$$

$$\geq \lambda(B(k_1,...,k_q))c_1 d + \lambda(B(k_1,...,k_q,k_{q+1}))$$

If we put $1 - c_1 d = c_3$, we have the result.
Multiplying each term of the sum we obtain

$$\sum_{k_{q+1} \in R} \lambda(B(k_1,..,k_q,k_{q+1}))^\sigma \geq \lambda(B(k_1,..,k_q))^\sigma c_3^{\sigma-1}(1 - c_2 \sum_{k_{q+1} \in I \setminus R} \lambda(B(k_{q+1}))$$

Here we must assume $c_2 \Sigma\lambda(B(k_{q+1})) < 1$ which is certainly true if R is not too small. We want to have

$$c_3^{\sigma-1}(1 - c_2 \sum_{k_{q+1} \in I \setminus R} \lambda(B(k_{q+1}))) = 1$$

This shows

$$\sigma = 1 - \frac{\log(1 - c_2 \Sigma\lambda(B(k_{q+1})))}{\log c_3} \geq 1 - c_4(R) \sum_{k_{q+1} \in I \setminus R} \lambda(B(k_{q+1}))$$

where $c_4(R)$ depends on the set R. On the other hand

$$\lambda(B(k_1,..,k_q)) = \sum_{k_{q+1} \in R} \lambda(B(k_1,..,k_q,k_{q+1})) + \sum_{k_{q+1} \in I \setminus R} \lambda(B(k_1,..,k_q,k_{q+1})) \geq$$

$$\geq \sum_{k_{q+1} \in R} \lambda(B(k_1,..,k_q,k_{q+1})) + c_1 \lambda(B(k_1,..,k_q)) \sum_{k_{q+1} \in (I \setminus R) \cap A} \lambda(B(k_{q+1}))$$

Here A denotes the set of all digits which can follow every sequence $(k_1,...,k_q)$.

Since

$$\sum_{k_{q+1} \in (I \setminus R) \cap A} \lambda(B(k_{q+1})) \geq c_0(R) \sum_{k_{q+1} \in I \setminus R} \lambda(B(k_{q+1}))$$

with some constant $c_0(R)$, $0 < c_0(R) < 1$ (see Schweiger [4] § 5 for similar combinatorial considerations) we finally obtain

$$\sum_{k_{q+1}\in R} \lambda(B(k_1,..,k_q,k_{q+1})) \le \lambda(B(k_1,..,k_q))(1 - c_1(R) \sum_{k_{q+1}\in I\setminus R} \lambda(B(k_{q+1}))$$

We note $\lambda(B(k_1,\dots,k_q,k_{q+1})) \ge c_1\lambda(B(k_1,\dots,k_q))M(R)$ where

$M(R) = \min_{k_{q+1}\in R} \lambda(B(k_{q+1}))$, and multiplying with

$$\lambda(B(k_1,\dots,k_q,k_{q+1}))^{\sigma-1} \le c_1^{\sigma-1} M(R)^{\sigma-1}\lambda(B(k_1,\dots,k_q))^{\sigma-1}$$

we obtain

$$\sum_{k_{q+1}\in R} \lambda(B(k_1,\dots,k_q,k_{q+1}))^{\sigma} \le \lambda(B(k_1,\dots,k_q))^{\sigma}[c_1 M(R)]^{\sigma-1} \cdot$$

$$\cdot \left[1 - c_1(R) \sum_{k_{q+1}\in I\setminus R} \lambda(B(k_{q+1}))\right]$$

We require $c_1^{\sigma-1}M(R)^{\sigma-1}\left[1 - c_1(R) \sum_{k_{q+1}\in I\setminus R} \lambda(B(k_{q+1}))\right] = 1$

This gives

$$\sigma = 1 - \frac{\log(1 - c_1(R) \sum_{k_{q+1}\in I\setminus R} \lambda(B(k_{q+1})))}{\log M(R) + \log c_1} \le 1 + c_5(R)\frac{\sum_{k_{q+1}\in I\setminus R} \lambda(B(k_{q+1}))}{\log M(R)}$$

<u>Corollary 18.4</u>: Let $M \ge M_o$ and $E_M = \{x \mid k_{sn}(x) \le M \text{ for } s \le 1\}$, then

$$1 - \frac{c_6}{M} \le \operatorname{bim} E_M \le 1 - \frac{c_7}{M \log M}$$

<u>Proof</u>: Since $k_{sj}(x) \le k_{sn}(x)$ for $i \le j \le n$, E_M is the set which can be described as a $E(R)$, where R is the set of digits the components of which are bounded by M. Since for $a = (a_1,\dots,a_n)$ $\lambda(B(a)) \sim a_n^{-(n+1)}$ the corollary follows from theorem 18.3 by routine estimations. One notes that for $M \ge M_o$ the constants c_6 and c_7 can be chosen uniformly.

<u>Corollary 18.5</u>: Let E be the set of points the digits of which are bounded, then $\operatorname{bim} E = 1$.

<u>Proof</u>: $\bigcup_{M\ge 1} E_M \subseteq E$

Without proof we mention (see Schweiger [8])

<u>Theorem 18.6</u>: Let $K \ge K_o$ and $F_K = \{x \mid k_{sn}(x) \ge K \text{ for } s \ge 1\}$, then

$$\frac{1}{2} + \frac{c_8}{2 \log K} \leq \text{bim } F_K \leq \frac{1}{2} + \frac{\log \log(K-1)}{2 \log(K-1)}$$

Theorem 18.7: Let $F = \{x \mid k_{sn}(x) \to \infty \text{ as } s \to \infty\}$. Then bim $F = \frac{1}{2}$

In connection with Good's paper we mention the paper by Hirst [1] .

§ 19. Ergodic invariant measures

We return to theorem 7.8. There was shown the following: Let

$$A(B(s_1,\ldots,s_t);\ m,x) = A(m,x) = \sum_{\substack{T^k x \in B(s_1,..,s_t) \\ 0 \leq k < m}} 1$$

for a given cylinder $B(s_1,\ldots,s_t)$, then $\lim_{m\to\infty} \frac{A(m,x)}{m} = \mu(B(s_1,..,s_t$

holds almost everywhere for any cylinder $B(s_1,\ldots,s_t)$.

Now let be P a measure such that $(B,\mathcal{L},P)$ is a probability space and T preserves P. The set of all T invariant probability measures P on $\mathcal{L}$ is denoted by I(T). An important subset from I(T) is the set EI(T)={P∈T(T is ergodic with respect to P}. A measure from EI(T) will be called ergodic.

A probability measure $P: \mathcal{L} \to R$ will be called an individual measure if there is an $x \in B$ such that

$$\lim_{m\to\infty} \frac{A(m,x)}{m} = P(B(s_1,\ldots,s_t))$$

for any cylinder $B(s_1,\ldots,s_t)$. In this case P is called the distribution measure of the sequence $(T^k x)_{k\in\mathbb{N}}$. The set of all individual measures is denoted by V(T).

Proposition 19.1: $V(T) \subseteq I(T)$

Proof: Since $\mathcal{V}^{(s)} \uparrow \mathcal{L}$ we will check invariance on cylinders only. Let $P \in V(T)$, then there exists an $x \in B$ such that

$$P(B(s_1,\ldots,s_t)) = \lim_{m\to\infty} \frac{A(m,x)}{m} .$$

Since $|A(m,x) - A(m,Tx)| \leq 1$ we obtain $\lim_{m\to\infty} \frac{A(m,x)}{m} = \lim_{\to\infty} \frac{A(m,Tx)}{m}$

Let $R \in I$ be any finite set of digits. Then

$$\sum_{a \in R} P(B(a,s_1,\dots,s_t)) = \sum_{a \in R} \lim_{m\to\infty} \frac{A(B(a,s_1,\dots,s_t);m,x)}{m} =$$

$$= \lim_{m\to\infty} \sum_{a \in R} \frac{A(B(a,s_1,\dots,s_t);m,x)}{m} \leq \lim_{m\to\infty} \sum_{a \in I} \frac{A(B(a,s_1,\dots,s_t);m,x)}{m} =$$

$$= \lim_{m\to\infty} \frac{A(T^{-1}B(s_1,\dots,s_t;m,x))}{m} = \lim_{m\to\infty} \frac{A(B(s_1,\dots,s_t);m,Tx)}{m} = P(B(s_1,\dots,s_t))$$

Since this inequality holds for all finite subsets $R \subseteq I$ we see

$$\sum_a P(B(a,s_1,\dots,s_t)) = P(T^{-1}B(s_1,\dots,s_t)) \leq P(B(s_1,\dots,s_t))$$

for any cylinder $B(s_1,\dots,s_t)$. Since $P(T^{-1}B) = 1$ we obtain

$$P(T^{-1}B(s_1,\dots,s_t)) = P(B(s_1,\dots,s_t))$$

Hence T preserves the measure P.

A theorem of J.Ville [1] (see A.G.Postnikov [1]) suggests the following conjecture: $V(T) = I(T)$. This theorem would be true if one can show that $V(T)$ is a closed convex subset of the set of all probability measures in the weak topology (see J.Cigler [1]).

Easy to prove is

Proposition 19.2: $EI(T) \subseteq V(T)$

Proof: Let $P \in EI(T)$, then

$$\lim_{m\to\infty} \frac{A(m,x)}{m} = P(B(s_1,\dots,s_t))$$

for P - almost all $x \in B$ according to the pointwise ergodic theorem.

To a given measure $P \in EI(T)$ we define the set $N(P) = \{x \mid P$ is the distribution measure of the sequence $(T^k x)_{k \in \mathbb{N}}\}$. Proposition 19.2 may be reparaphrased as $P[N(P)] = 1$

Taking $P = \mu$, we have theorem 7.8 . Since λ and μ are equivalent, we can state

$$\lambda[N(\mu)] = 1 \quad \text{or equivalently}$$
$$\lambda[B \setminus N(\mu)] = 0$$

Theorem 19.3: $\dim [B \setminus N(\mu)] = 1$

Proof: Note that $E \subseteq B \setminus N(\mu)$, if E denotes the set of points the digits of which are bounded (see corollary.18.5).

In this connection we mention the related problem of "normality with respect to a number-theoretical transformation" , see Schweiger [12] (the proof of Hilfssatz 3 is not correct, but the result follows fro a theorem by Cigler [1] (Satz 7); Hilfssatz 5 should be completed b using a theorem'similar to theorem 17.1 as given by Schweiger-Stradne [1], [2]).

If $P \neq \mu$, we have $\lambda[N(P)] = 0$

One of the most difficult questions not even completely solved for co tinued fractions is the calculation of bim N(P). We refer the interested reader to the papers by J.R. Kinney-T.S.Pitcher [1], and Schwei ger [11] . The case of g-adic fractions was extensively studied by B.Volkmann [1] and C.M.Colebrook [1]. In all these one dimensional cases one has results for dim N(P) too. We only sketch the proof of

<u>Theorem 19.4</u>: If the function $\log x_1$ is integrable with respect to P, then

$$\lim_{t\to\infty} \frac{P(B(s_1,\ldots,s_t))}{\lambda(B(s_1,\ldots,s_t))} = \alpha$$

exists for P-almost all $x = \Psi[s_1,\ldots,s_t,\ldots]$ and bim $[N(P)] \geq$

<u>Proof</u>: If $\log x_1$ is P-integrable then $\log D(x) = -(n+1) \log x_1$ is P-integrable too. Then similarly to theorem 7.9

$$\lim_{t\to\infty} \frac{1}{t} \log \lambda(B(s_1,\ldots,s_t)) = -h_o(T)$$

exists on a set W_1 with $P(W_1) = 1$. The theorem of McMillan (see Billingsley [1] or Parry [1]) shows

$$\lim_{t\to\infty} \frac{1}{t} \log P(B(s_1,\ldots,s_t)) = -h_P(T)$$

exists on a set W_2 with $P(W_2) = 1$. The proof of this theorem conceptually involves information theory which will not be presented here.

Putting $\alpha = \frac{h_P(T)}{h_o(T)}$ and using $N(P) \cap W_1 \cap W_2 \subseteq N(P)$ we get the resul (note that $P[N(P) \cap W_1 \cap W_2] = 1$) with help of theorem 16.3 .

<u>Proposition 19.5</u>: Let $\nu \in I(T)$, $P \in EI(T)$ and ν be absolutely con - tinuous with respect to P, then $\nu = P$.

Proof: The assumption, ν absolutely continuous with respect to P, means: $P(M) = 0$ implies $\nu(M) = 0$. Let $A \in \mathcal{L}$ be arbitrary and c_A its indicator function. Then

$$\lim_{N\to\infty} \frac{1}{N} \sum_{j=0}^{N-1} c_A (T^j x) = P(A)$$

on a set K_1 with $P(B \setminus K_1) = 0$. Hence $\nu(B \setminus K_1) = 0$ and $\nu(K_1) = 1$. On the other hand

$$\lim_{N\to\infty} \frac{1}{N} \sum_{j=0}^{N-1} c_A(T^j x) = E(c_A \parallel \mathcal{Y})$$

on a set K_2 with $\nu(K_2) = 1$. Hence $E(c_A \parallel \mathcal{Y}) = P(A)$ ν-almost everywhere. Since $\int_B c_A \, d\nu = \int_B E(c_A \parallel \mathcal{Y}) d\nu$ we obtain $\nu(A) = P(A)$.

Corollary 19.6: Let $\nu \in I(T)$ and ν absolutely continuous with respect to μ, then $\nu = \mu$.

As usually we call a measure P singular with respect to λ, if there is a set $E \subseteq B$ such that $\lambda(E) = 1$, but $P(E) = 0$.

Proposition 19.7: If $P \in EI(T)$ and $P \neq \mu$, then P is singular with respect to λ .

Proof: Take $E = N(\mu)$, then $\lambda(E) = 1$, but $P(N(\mu)) = 0$.

With the help of martingal theory (see Doob [1]) or the theory of differentiation with respect to nets (see Munroe [1]) one could prove

Theorem 19.8: Let $P \in EI(T)$, then P is singular if and only if

$$\lim_{t\to\infty} \frac{P(B(s_1,..,s_t))}{\lambda(B(s_1,..,s_t))} = 0 \quad \text{for } \lambda\text{-almost all} \quad x = \Psi\,[s_1, s_2, \ldots] .$$

A very interesting question is suggested by a paper of Smorodinsky [1]: Let $P \in EI(T)$. Can there be selected a set $M \subseteq B$ and a constant $c > 0$, such that $P(M) = 1$, $P(A) = cH^*(\gamma, \mathfrak{Z}; A \cap M)$ where $H^*(\gamma, \mathfrak{Z}; .)$ is the γ-dimensional Hausdorff measure, induced by the semimetric given by the cylinders ? Smorodinsky gives some results for 2-adic expansions, Jakubec [1] , [3] for Lüroth expansions. Theorem 19.4 shows, that $\gamma = \alpha$ (defined in this very theorem) is the only possible value of γ .

The proof of the extension to continuous concave functions Ψ in Smorodinsky [1] contains a gap.

Another question is the following: Let P be an ergodic, invariant measure with a Markovian dependence of length n-1 (n is the dimension of B), namely

$$\frac{P(B(a_1,\dots,a_s,\ b_1,\dots,b_{n-1}))}{P(B(a_1,\dots,a_s))} = p(b_1,\dots,b_{n-1})$$

The following conjecture was made: Such a measure is singular. The conjecture was proved for n=1 by Chatterji [1] and Schweiger [1], for n=2 by Schweiger [5]. It would be interesting to have a proof for $n \geq 3$. For Lüroth expansions Jakubec [2] has proven some interesting results.

§ 2o. Volume as approximation measure

The results of § 9 have shown that approximation questions are in fact very difficult to handle. The differences

$$\alpha_j - \frac{A_j^{(s+n)}}{A_o^{(s+n)}}, \quad 1 \leq j \leq n$$

do not easy generalize the properties of continued fractions (n=1). It is remarkable that several properties of continued fractions have a more friendly behaviour if we generalize in another way. Let $\alpha = (\alpha_1,\dots,\alpha_n) \in B$ be given and

$$p^{(s+j)} = \left(\frac{A_1^{(s+j)}}{A_o^{(s+j)}},\dots,\frac{A_n^{(s+j)}}{A_o^{(s+j)}}\right), \quad 1 \leq j \leq n \quad \text{will denote n consecutive}$$

approximation points to α. These n+1 points span a convex polyhedron $P(\alpha;s)$. We now put $V(\alpha;s) = \lambda(P(\alpha;s))$

Elementary analytic geometry shows

$$V(\alpha;s) = \left| \frac{1}{n!} \det\left(\left(\alpha_i - \frac{A_i^{(s+j)}}{A_o^{(s+j)}}\right)\right) \right|, \quad 1 \leq i, j \leq n$$

By the calculation made in lemma 2.4 we see

$$V(\alpha;s) = \left[n!\, A_o^{(s+1)} \dots A_o^{(s+n)}\right]^{-1} \det((\alpha_i A_o^{(+j)} - A_i^{(s+j)})) =$$

$$= \frac{1}{n!\, A_o^{(s+1)} \dots A_o^{(s+n)} (A_o^{(s+n+1)} + \sum_{j=1}^{n} A_o^{(s+j)} \eta_j)} \quad \text{where} \quad \eta = T^s \alpha .$$

This gives an elementary estimate:

Proposition 2o.1:

$$\frac{1}{A_o^{(s+n+1)} + \sum_{j=1}^{n} A_o^{(s+j)}} \leq n!\, A_o^{(s+1)} \ldots A_o^{(s+n)}\, V(\alpha;s) \leq \frac{1}{A_o^{(s+n+1)}}$$

Proof: Note, that $0 \leq \eta_j \leq 1$.

Proposition 2o.2:

$$\lim_{s\to\infty} \frac{\log(V(x;s)}{s} = - h(T) \qquad \lambda - \text{a.e.}$$

Proof: By proposition 2o.1 we have (using $A_o^{(s+n+1)} \geq A_o^{(s+j)}$, $1 \leq j \leq n$)

$$\frac{1}{(n+1)A_o^{(s+n+1)}} \leq n!\, A_o^{(s+1)} \ldots A_o^{(s+n)}\, V(x;s) \leq \frac{1}{A_o^{(s+n+1)}}$$

or

$$- \log(n+1) - \log A_o^{(s+n+1)} \leq \log n! + \sum_{j=1}^{n} \log A_o^{(s+j)} + \log V(x;s) \leq$$

$$\leq - \log A_o^{(s+n+1)}$$

Dividing by s and using corollary 7.1o we obtain the result .

One introduces the quantities

$$F_j(\alpha;s) = \frac{A_o^{(s+n+1)} + \sum_{i=1}^{n} A_o^{(s+i)}\eta_i}{A_o^{(s+j)}} \quad , 1 \leq j \leq n$$

and obtains the estimates

$$n!\, A_o^{(s+1)} \ldots A_o^{(s+n)}\, V(\alpha;s) \leq \frac{1}{A_o^{(s+j)}\, F_j(\alpha;s)}$$

The main interest lies in the consideration of

$$\limsup_{s\to\infty} \frac{1}{F_j(\alpha;s)} = K_j(\alpha) \quad \text{and} \quad \liminf_{s\to\infty} \frac{1}{F_j(\alpha;s)} = \kappa_j(\alpha)$$

Clearly $\kappa_j(\alpha) \leq K_j(\alpha) \leq 1$ for $1 \leq j \leq n$. Since $A_o^{(s+1)} \leq \ldots \leq A_o^{(s+n)}$ we obviously have $K_1(x) \leq \ldots \leq K_n(x)$ and $\kappa_1(x) \leq \ldots \leq \kappa_n(x)$

We now prove

Theorem 2o.3: $\inf_{x\,B} \kappa_j(x) = 0, \quad 1 \leq j \leq n$

Proof: We will prove the existence of $x \in B$ such that $\kappa_n(x) = 0$

Now

$$F_n(x;s) \geq \frac{A_o^{(s+n+1)}}{A_o^{(s+n)}} \geq k_{sn}(x)$$

Taking any x with $\limsup_{s \to \infty} k_{sn}(x) = \infty$, we obtain the result. Note that λ-almost all $x \in B$ have this property.

On the other hand we have

Theorem 2o.4: $\sup_{x \in B} K_j(x) = 1$, $1 \leq j \leq n$

Proof: We will show: $K_1(x) = 1$ for some $x \in B$.
We start with $(T^s x = y)$:

$$F_1(x;s) = \frac{A_o^{(s+n+1)} + \sum_{j=1}^{n} A_o^{(s+j)} y_j}{A_o^{(s+1)}} \leq \frac{A_o^{(s+n+1)}(1 + \sum_{j=1}^{n} y_j}{A_o^{(s+1)}}$$

Now we choose $k_j = (0,0,\ldots,0,1)$ for $s-n+1 \leq j \leq s$ (if n=1, we choose $k_s = 1$)

Then

$$A_o^{(s+n+1)} = A_o^{(s+n)} + A_o^{(s)} = A_o^{(s+n-1)} + A_o^{(s)} + A_o^{(s-1)} = \ldots$$

$$\ldots = A_o^{(s+1)} + A_o^{(s)} + \ldots + A_o^{(s-n+1)}$$

Hence

$$\frac{A_o^{(s+n+1)}}{A_o^{(s+1)}} \leq 1 + \sum_{i=0}^{n-1} \frac{A_o^{(s+i)}}{A_o^{(s+1)}} \leq 1 + n\frac{A_o^{(s)}}{A_o^{(s+1)}}$$

Choosing M(s) sufficiently great and $k_{s-n} = (0,\ldots,0,M(s))$

we see

$$\frac{A_o^{(s+n+1)}}{A_o^{(s+1)}} \leq 1 + \frac{n}{M(s)} \leq 1 + \frac{1}{s}$$

We now choose the digits k_j for $s+1 \leq j \leq s+L(s)$ such that

$$1 + \sum_{j=1}^{n} y_j \leq 1 + \frac{1}{s}$$

and obtain $F_1(x;s) \leq 1 + \frac{3}{s}$

We now construct a sequence $[k_1, k_2, ..]$ in doing the described choice for a sequence of indices $s=s_k$ (starting with $s_1 = n+1$ and recursively defined by $s_{n+1} = s_n + n + 1 + L(s))$. Then $\lim_{k\to\infty} F_1(x;s_k) = 1$, which gives the result.

This theorem goes back to a device of R. Fischer.

Proposition 2o.5: $\inf_{x \in B} K_j(x) = 0, \quad 1 \leq j \leq n$

Proof: We will prove the existence of $x \in B$ such that $K_n(x) = 0$. Since $F_n(x;s) \geq k_{sn}(x)$ we have only to take an x such that $(k_{sn}(x))_{s \in \mathbb{N}}$ increases monotonically to infinity. Then $\lim_{s\to\infty} \frac{1}{F_n(x;s)} = 0$

Now we are left with the most difficult problem, namely to determine $\sup_{x \in B} \kappa_j(x)$. In fact the problem to determine $\kappa_j(x)$ for various classes of x is one of the most interesting, but nearly unsolved problems. Only for n=1 (continued fractions) the detailed theory of the Markoff spectrum is available (see Cassels [1] , Cohn [1], Marshall Hall [1]). Since $\kappa_j(x) = 0$ for λ-almost all x the problem is more "arithmetical" than metrical. The inequality

$$\frac{A_o^{(s+n+1)}}{A_o^{(s+n)}} \leq F_n(x;s) \leq F_1(x;s) \leq \frac{\sum_{i=1}^{n+1} A_o^{(s+i)}}{A_o^{(s+1)}} \leq \frac{(n+1)A_o^{(s+n+1)}}{A_o^{(s+n)}}$$

gives $F_1(x;s) \leq (n+1)F_n(x;s)$ and $k_{sn} \leq F_n(x;s) \leq F_1(x;s) \leq (n+1)^2 k_{sn}$

Therefore $\kappa_n(x) > 0 \iff \kappa_1(x) > 0$ and $\kappa_n(x) > 0$ if and only if the set $\{k_{sn}\}$ is bounded.

The following conjecture seems reasonable: Let ξ be the unique root of $\xi^{n+1} - \xi^n - 1 = 0$ with $\xi > 1$, then

$$\sup_{x\ B} \kappa_j(x) = \frac{\xi^j}{\xi^n + n + 1}, \quad 1 \leq j \leq n$$

Moreover, for every x B the inequality $F_j(x;s) \quad \frac{\xi^n + n + 1}{\xi^j}$ is satisfied for infinitely many s. For n=1 (continued fractions) we obtain

$$\xi = \frac{1 + \ 5}{2} \qquad \sup_{x\ B} \kappa_1(x) = \frac{1}{\sqrt{5}}$$

which is just Hurwitz' theorem (see Khinchine [1] , Cassels [1]).

<u>Proposition 2o.6</u>: Let α given and $T^h\alpha = \gamma$, $\gamma = (\xi^{-1}, \xi^{-2},..,\xi^{-n}) \in B$, then $F_j(\alpha;s) > \frac{\xi^n + n + 1}{\xi^j}$ and $F_j(\alpha;s) < \frac{\xi^n + n + 1}{\xi^j}$ are both soluble for infinitely many s.

<u>Proof</u>: We put $\beta_j = \frac{\xi^n + n + 1}{\xi^j}$ and note that $\gamma = (\xi^{-1},..,\xi^{-n})$ has the purely periodic expansion $k_s(\gamma) = (0,..,0,1)$, $s \geq 1$.

We introduce the hyperplane defined by the equation

$$L(x_1,x_2,..,x_n) = x_1 + x_2\xi^{-n} + x_3\xi^{-n+1} + \dots + x_{n-j+1}\xi^{-j-1} +$$

$$+ x_{n-j+2}\xi^{-j+1} + \dots + x_{n-1}\xi^{-2} + x_n\xi^{-1} + \xi^{-j} - \beta_j = 0$$

By the choice of β_j the point $\Theta = (\xi^{n+1-j}, \xi^{n-j},\dots,\xi,\xi^{-1},\dots,\xi^{-j+1})$ lies on the hyperplane. In fact $L(\Theta) = 0$ follows by an easy calculation. Now we introduce the points

$$p(s) = \left(\frac{A_0^{(s+n+1)}}{A_0^{(s+j)}}, \dots, \frac{A_0^{(s+j+1)}}{A_0^{(s+j)}}, \frac{A_0^{(s+j-1)}}{A_0^{(s+j)}}, \dots, \frac{A_0^{(s+1)}}{A_0^{(s+j)}}\right)$$

The recursion formula

$$A_0^{(g+n+1)} = A_0^{(g+n)} + A_0^{(g)} , \quad g \geq h$$

shows the following:

(1) $p(s+n+1)$ lies in the convex polyhedron spanned by the n+1 points $p(s),\dots,p(s+n)$ (if $s \geq h$)

(2) $\lim_{s\to\infty} p(s) = \Theta$

(3) Θ lies in each convex polyhedron spanned by $p(s),\dots,p(s+n)$ for $s \geq h$

Since $F(x) = x^{n+1} - x^n - 1$ is irreducibel we have $L(p(s)) \neq 0$ for every $s \geq h$. If $p(s),\dots,p(s+n)$ are n+1 subsequent points, at least one of them satisfies $L(p(\nu)) < 0$, $s \leq \nu \leq s+n$ and at least one other point $L(p(\mu)) > 0$, $s \leq \mu \leq s+n$

This shows the assertion of the theorem true. Moreover we see

$$\lim_{s\to\infty} F_j(\alpha;s) = \beta_j$$

<u>Corollary 2o.7</u>: The conjecture would be proved, if one could show $F_j(x;s) > \beta_j$ is soluble for infinitely many s for any $x \subseteq B$.

§ 21. Proof of the conjecture for n=1 and n=2

Although the case n=1 is well known and the theory of the Markoff spectrum gives in fact much more and deeper information, we will present a proof for n=1 to motivate the technique used for n=2. This technique goes back to an important paper of W.M.Schmidt [1] .

Theorem 21.1: Let n=1 . Then $F_1(x;s) > \sqrt{5}$ has infinitely many solutions s for any $x \in [0,1]$.

Proof: We recall

$$F(x;s) = \frac{A_o^{(s+2)} + A_o^{(s+1)} x^{(s)}}{A_o^{(s+1)}} , \quad F(x;s+1) = \frac{A_o^{(s+3)} + A_o^{(s+2)} x^{(s+1)}}{A_o^{(s+2)}}$$

where $x(s) = T^s x$ etc. Since

$$A_o^{(s+2)} = k_1^{(s)} A_o^{(s+1)} + A_o^{(s)} , \quad A_o^{(s+3)} = k_1^{(s+1)} A_o^{(s+2)} + A_o^{(s+1)}$$

we have

$$F(x;s) \geq k_1^{(s)} \qquad F(x;s+1) \geq k_1^{(s+1)}$$

Now suppose that there is a t_o such that $F(x;t) \leq \sqrt{5}$ for all $t \geq t_o$ then $k_1^{(s)} \leq 2$ and $k_1^{(s+1)} \leq 2$. If there is an $s \geq t_o$, such that $k_1^{(s)}=2$, then

$$5 \geq F(x;s) = \frac{2A_o^{(s+1)} + A_o^{(s)} + A_o^{(s+1)} x^{(s)}}{A_o^{(s+1)}}$$

shows $2 + x^{(s)} \leq 5$

Since $x^{(s)} = \dfrac{1}{k_1^{(s+1)} + x^{(s+1)}} \geq \dfrac{1}{k_1^{(s+1)} + 1}$ we arrive at $2 + \frac{1}{3} \leq 5$

which is clearly a contradiction. Therefore $k_1^{(s)} = 1$ for all $s \geq t_o$ and proposition 2o.6 shows the theorem true.

Theorem 21.2: Let n = 2. Then infinitely often $F_1(x;s) > \beta_1$ where

$$\beta_1 = \frac{\xi^2 + 3}{\xi} = 3\xi^2 - 2\xi , \quad \xi > 1 \text{ and } \xi^3 = \xi^2 + 1.$$

Proof: We begin with

$$F_1(x;s+1) = \frac{A_o^{(s+4)} + A_o^{(s+3)} x_2^{(s+1)} + A_o^{(s+2)} x_1^{(s+1)}}{A_o^{(s+2)}}$$

$$F_1(x;s+2) = \frac{A_o^{(s+5)} + A_o^{(s+4)}x_2^{(s+2)} + A_o^{(s+3)}x_1^{(s+2)}}{A_o^{(s+3)}}$$

We note that $\beta_1^3 - \beta_1^2 - 31 = 0$ and therefore $3 < \beta_1 < 4$

We introduce the quantities

$$L = \frac{A_o^{(s+3)}}{A_o^{(s+2)}} \quad , \quad M = \frac{A_o^{(s+2)}}{A_o^{(s+1)}}$$

$$\lambda = x_2^{(s+2)} + k_2^{(s+2)} \qquad \mu = x_2^{(s+3)} + k_2^{(s+3)}$$

$$a = k_2^{(s+1)} \qquad b_1 = k_1^{(s+1)} \qquad b_2 = k_1^{(s+2)}$$

and use the formulas

$$x_2^{(s+1)} = \frac{x_1^{(s+2)} + k_1^{(s+2)}}{x_2^{(s+2)} + k_2^{(s+2)}}$$

$$x_1^{(s+1)} = \frac{1}{x_2^{(s+2)} + k_2^{(s+2)}}$$

$$x_1^{(s+2)} = \frac{1}{x_2^{(s+3)} + k_2^{(s+3)}}$$

$$A_o^{(s+4)} = A_o^{(s+1)} + A_o^{(s+2)}k_1^{(s+1)} + A_o^{(s+3)}k_2^{(s+1)}$$

Comparing $F_1(x;s+1)$ with $F_2(x;s+2)$ and developping $F_1(x;s+1)$ into the new quantities we obtain $F_1(x;s+1) = F_1(x;s+2)\frac{L}{\lambda}$

$$F_1(x;s+1) = \frac{1}{M} + \frac{1}{\lambda} + b_1 + L(a + \frac{1+\mu b_2}{\mu\lambda})$$

We introduce

$$k = \mu\lambda M\, F_1(x;s+1) = M\mu + LM(b_2\mu+1) + \lambda\mu(aLM+b_1M+1) = \mu M(1+Lb_2+\lambda b_1+aL\lambda)$$
$$+ LM + \lambda\mu$$

and we obtain the equations

$$k - \mu\lambda M\, F_1(x;s+1) = 0$$

$$k - L\mu M\, F_1(x;s+2) = 0$$

Now let $F_1(x;t) \leq \beta_1$ for all $t \geq t_o$. Hence

$$F_1(x;s+1) = \varepsilon_1\beta_1$$

$$F_1(x;s+2) = \varepsilon_2\beta_1$$

with $0 < \varepsilon_1, \varepsilon_2 \le 1$. We obtain the equations

$$k - \mu\lambda M\varepsilon_1\beta_1 = 0$$

$$k - \mu LM\varepsilon_2\beta_1 = 0$$

In particular: $\lambda\varepsilon_1 = L\varepsilon_2$. The second equation now gives:

$$\mu M(1 + Lb_2 + \lambda b_1 + aL\lambda) + LM + \lambda\mu - \varepsilon_2\mu LM\beta_1 = 0$$

$$\varepsilon_2\mu LM\left(\frac{1}{\varepsilon_2 L} + \frac{b_2}{\varepsilon_2} + \frac{\lambda b_1}{L\varepsilon_2} + \frac{a\lambda}{\varepsilon_2} - \beta_1\right) + LM + \lambda\mu = 0$$

Now

$$\frac{b_2}{\varepsilon_2} + \frac{\lambda b_1}{L\varepsilon_2} = \frac{b_2}{\varepsilon_2} + \frac{b_1}{\varepsilon_1} \ge b_2 + b_1$$

$$\frac{1}{\varepsilon_2 L} + \frac{a\lambda}{\varepsilon_2} = \frac{1}{\varepsilon_2 L} + \frac{aL}{\varepsilon_1} \ge \frac{1}{L} + aL \ge (a-1) + L + \frac{1}{L} \ge a + 1$$

Therefore

$$a + b_1 + b_2 + 1 - \beta_1 \le 0$$

Since $\beta_1 < 4$ we obtain $a + b_1 + b_2 \le 2$

or

$$k_2^{(s+1)} + k_1^{(s+1)} + k_1^{(s+2)} \le 2 \qquad \text{for all } s \ge t_o.$$

If $k_1^{(s+1)} = 1$, then $k_2^{(s+1)} = 1$ and by the restrictions on digits $k_1^{(s+2)} = 1$, which is impossible. Therefore $k_1^{(s)} = 0$ for all $s \ge t_o + 1$. We have only to look at the various possibilities given by $k_2^{(s)}$. If $k_2^{(s)} = 1$ for all $s \ge s_o$ we use proposition 20.6. We now assume $k_2^{(s+2)} = 2$ and using $A_o^{(s+5)} = 2A_o^{(s+4)} + A_o^{(s+2)} = 2k_2^{(s+1)}A_o^{(s+3)} + 2A_o^{(s+1)} + A_o^{(s+2)}$ in this case, we obtain

$$F_1(x;s+2) \ge 2k_2^{(s+1)} + \frac{A_o^{(s+2)} + 2A_o^{(s+1)}}{A_o^{(s+3)}} + x_1^{(s+2)} + x_2^{(s+2)}$$

We distinguish the cases:

(a) $k_2^{(s+1)} = 2$

Then $F_1(x;s+2) \geq 4$

In cases (b) and (c) we have $x_1^{(s+2)} + x_2^{(s+2)} \geq 1$

(b) $k_2^{(s+1)} = 1,\ k_2^{(s)} = 1$

Then $A_o^{(s+3)} = A_o^{(s+2)} + A_o^{(s)}$ and

$$F_1(x;s+2) \geq 3 + \frac{A_o^{(s+2)} + 2A_o^{(s+1)}}{A_o^{(s+2)} + A_o^{(s)}} \geq 4$$

(c) $k_2^{(s+1)} = 1,\ k_2^{(s)} = 2,\ k_2^{(s-1)} = 1,\ k_2^{(s-2)} = 2$.

Then

$$\frac{A_o^{(s+2)} + 2A_o^{(s+1)}}{A_o^{(s+3)}} = \frac{A_o^{(s+2)} + 2A_o^{(s+1)}}{2A_o^{(s+2)} + A_o^{(s)}} = \frac{3A_o^{(s+1)} + A_o^{(s-1)}}{2A_o^{(s+1)} + 2A_o^{(s-1)} + A_o^{(s)}} =$$

$$= \frac{6A_o^{(s)} + A_o^{(s-1)} + 3A_o^{(s-2)}}{5A_o^{(s)} + 2A_o^{(s-1)} + 2A_o^{(s-2)}} \geq \frac{5A_o^{(s)} + 2A_o^{(s-1)} + 3A_o^{(s-2)}}{5A_o^{(s)} + 2A_o^{(s-1)} + 2A_o^{(s-2}} \geq 1$$

Hence $F_1(x;s+2) \geq 4$.

One sees at once that if $k_2^{(s+2)} = 2$ infinitely often case (a), (b) or (c) must hold which all lead to a contradiction. Henceforth the theorem is proved.

W.M.Schmidt [1] uses a more sophisticated technique showing that in fact $F_1(x;s) > \frac{13}{3}$ infinitely often with the exception of one further case, namely, when the algorithm becomes purely periodic in the following manner: The period has length two and $k_2^{(N+2n)} = 1,\ k_2^{(N+2n+1)} =$ $k_1^{(N+n)} = 0$ for $n \geq 1$ with suitable $N \geq 1$. In this case

$$\lim_{s\to\infty} F_1(x;s) = \xi \quad \text{where } \xi^3 - 3\xi^2 - 23 = 0 \text{ and } \xi > 1 .$$

<u>Theorem 21.3</u>: Let $n = 2$. Then infinitely often $F_2(x;s) > \beta_2$ where $\beta_2 = \frac{\xi^2 + 3}{\xi^2} = 3\xi - 2$, $\xi > 1$, and $\xi^3 = \xi^2 + 1$.

<u>Proof</u>: We note $\beta_2^3 + 3\beta_2^2 - 31 = 0$ and therefore $2 < \beta_2 < \frac{29}{12}$

We start with

$$F_2(x;s+2) = k_2^{(s+2)} + x_2^{(s+2)} + \frac{A_o^{(s+3)}}{A_o^{(s+4)}}(k_1^{(s+2)} + x_1^{(s+2)}) + \frac{A_o^{(s+2)}}{A_o^{(s+4)}}$$

If $F_2(x;t) \leq \beta_2$ for $t \geq t_o$, then clearly $k_2^{(t)} \leq 2$. We list several cases.

(a) $k_2^{(t)} = 2$ infinitely often. We can assume $k_2^{(s+2)} = 2$.

(a.1) $k_1^{(s+2)} \geq 1$

We note

$$x_2^{(s+2)} = \frac{k_1^{(s+3)} + x_1^{(s+3)}}{k_2^{(s+3)} + x_2^{(s+3)}} \geq \frac{1}{9}$$

$$x_1^{(s+2)} = \frac{1}{k_2^{(s+3)} + x_2^{(s+3)}} \geq \frac{1}{3}$$

$$A_o^{(s+4)} \leq 2A_o^{(s+3)} + 3A_o^{(s+2)}$$

Then

$$F_2(x;s+2) \geq 2 + \frac{1}{9} + \frac{A_o^{(s+3)}}{A_o^{(s+4)}}(1 + \frac{1}{3}) + \frac{A_o^{(s+2)}}{A_o^{(s+4)}} \geq \frac{19}{9} + \frac{4A_o^{(s+3)} + 3A_o^{(s+2)}}{3A_o^{(s+4)}} \geq$$

$$\geq \frac{19}{9} + \frac{4A_o^{(s+3)} + 3A_o^{(s+2)}}{6A_o^{(s+3)} + 9A_o^{(s+2)}} \geq \frac{19}{9} + \frac{1}{3} = \frac{22}{9} > \frac{29}{12}$$

(a.2) $k_1^{(s+2)} = 0$

We first assume that the algorithm becomes periodic in the form $k_2^{(t)} = 2$, $k_1^{(t)} = 0$ for all $t \geq t_1 \geq t_o$. In this case $T^t 1x = \gamma$ where

$$\gamma = (\theta^{-1}, \theta^{-2}), \quad \theta > 1 \qquad \text{and} \qquad \theta^3 - 2\theta^2 - 1 = 0$$

In this case $\lim_{s\to\infty} F_2(x;s) = 3\theta - 4 > \beta_2$

If this situation does not enter we assume:

$$k_2^{(s+2)} = 2, \quad k_1^{(s+2)} = 0, \quad k_2^{(s+1)} = 1, \quad k_1^{(s+1)} \leq 1$$

Then again $x_2^{(s+2)} \geq \frac{1}{9}$, $x_1^{(s+2)} \geq \frac{1}{3}$ but $A_o^{(s+4)} \leq A_o^{(s+3)} + 2A_o^{(s+2)}$

We estimate

$$F_2(x;s+2) \geq 2 + \frac{1}{9} + \frac{A_o^{(s+3)}}{A_o^{(s+4)}} \cdot \frac{1}{3} + \frac{A_o^{(s+2)}}{A_o^{(s+4)}} \geq \frac{19}{9} + \frac{A_o^{(s+3)} + 3A_o^{(s+2)}}{3A_o^{(s+4)}} \geq$$

$$\geq \frac{19}{9} + \frac{A_o^{(s+3)} + 3A_o^{(s+2)}}{3A_o^{(s+3)} + 6A_o^{(s+2)}} \geq \frac{19}{9} + \frac{1}{3} = \frac{22}{9} > \frac{29}{12}$$

(b) $k_2^{(t)} = 1$ for all $t \geq t_o$. If $k_1^{(t)} = 0$ for all $t \geq t_o$, proposition 2o.6 applies. If $k_1^{(t)} = 1$ for any $t_1 \geq t_o$, then the admissibility conditions imply $k_1^{(t)} = 1$ for all $t \geq t_1$. In this case $T^{t_1} x = \gamma$ where $\gamma = (\theta^{-1}, \theta^{-1} + \theta^{-2})$, $\theta > 1$ and $\theta^3 - \theta^2 - \theta - 1 = 0$. In this case $\lim_{s \to \infty} F_2(x;s) = 2\theta - 1 + \theta^{-2} > \beta_2$.

The extensions of theorem 21.2 to n = 3 and n = 4 will appear in forth coming papers by Schweiger in Math.Zeitschrift and Acta Arithmetica.

Appendix I

I want to point out the connection with Perron' s notation [1] :

Perron does not restrict his attention to the unit cube, therefore we could define more generally

$$\Lambda_o = \begin{pmatrix} 0 & 0 & \dots & 0 & 1 \\ 1 & 0 & \dots & 0 & a_{o1} \\ . & . & . & . & . \\ 0 & 0 & \dots & 1 & a_{on} \end{pmatrix}$$

and $A_i^{(n+1)} = a_{oi} \quad , 1 \leq i \leq n$.

Restriction to the unit cube corresponds to $a_{oi} = 0, 1 \leq i \leq n$.

Perron starts with $(\alpha_1,\dots,\alpha_n)$ and using the notation $\left[\alpha_i^{(s)}\right] = a_i^{(s)}$

he puts

$$\alpha_1^{(s)} = a_1^{(s)} + \frac{1}{\alpha_n^{(s+1)}} \ , \ \alpha_2^{(s)} = a_2^{(s)} + \frac{\alpha_1^{(s+1)}}{\alpha_n^{(s+1)}} \ , \ \dots$$

$$\dots \ , \ \alpha_n^{(s)} = a_n^{(s)} + \frac{\alpha_{n-1}^{(s+1)}}{\alpha_n^{(s+1)}} \ .$$

The correspondence is the following: $T^s x = y \quad , s \geq 0$.

$$y_1 = \frac{1}{\alpha_n^{(s+1)}} \ , \ y_2 = \frac{\alpha_1^{(s+1)}}{\alpha_n^{(s+1)}} \ , \ \dots \ , \ y_n = \frac{\alpha_{n-1}^{(s+1)}}{\alpha_n^{(s+1)}}$$

$$k_{si}(x) = a_i^{(s)} \ .$$

Appendix II

Metrical results on a p-adic Jacobi algorithm by A.A.Ruban [1]. Jacobi algorithm in fields of formal power series is considered by R.Paysant Le Roux - E. Dubois [1].

References

/1/ Bernstein,L.: The Jacobi-Perron Algorithm. Its Theory and Application. Lecture Notes in Mathematics 2o7. Springer 1971.

/1/ Billingsley,P.: Ergodic Theory and Information. J.Wiley, New York, 1965.

/2/ Billingsley,P.: Hausdorffdimension in Probability Theory. Ill.J. Math. 4 (196o), 187-2o9.

/3/ Billingsley,P.: Hausdorffdimension in Probability Theory II. Ill.J.Math. 5 (1961), 291-298.

/1/ Bottorf,G.A.: Proof of the Littlewood Conjecture for Infinitely Many Pairs. Notices AMS 19 , A-622.

/1/ Caldéròn,A.P.: Sur les mesures invariantes, C.R.Acad.Sci., Paris 24o (1955), 196o-1962.

/1/ Cassels,J.W.S.: An Introduction to Diophantine Approximation, Cambridge Tracts No. 45 (1957).

/1/ Chatterji,S.D.: Maße, die von regelmäßigen Kettenbrüchen induziert sind, Math.Ann. 164 (1966), 113-117.

/1/ Cigler,J.: Der individuelle Ergodensatz in der Theorie der Gleichverteilung mod 1, J.Reine Angew.Math. 2o5 (196o/61), 91-1oo.

/1/ Cohn,H.: Representation of Markoff's Binary Quadratic Forms by Geodesics on a Perforated Torus, Acta Arithm. 18 (1971), 125-136.

/1/ Colebrook,C.M.: The Hausdorff Dimension of Certain Sets of Nonnormal Numbers, Mich.Math.J. 17 (197o) , 1o3-1o6.

/1/ Doob,J.L.: Stochastic Processes, Wiley, New York 1953.

/1/ Ennola,V.: On Metric Diophantine Approximation, Ann.Univ.Turku Ser. AI 113, (1967), 1-7

/1/ Federer,H.: Geometric Measure Theory. Die Grundlagen der math. Wiss. 153, Springer, Berlin 1969.

/1/ Fischer,R.: Konvergenzgeschwindigkeit beim Jacobialgorithmus. Anzeiger der math.-naturw.Klasse der Österreichischen Akademie der Wissenschaften.(to appear)

/1/ Foguel,S.R.: The Ergodic Theory of Markov Processes. Van Nostrand, London-New York 1969.

/1/ Friedmann,N.A.: Introduction to Ergodic Theory. Van Nostrand Reinhold Comp. 197o.

/1/ Gaal,St.A.: Point Set Topology. Academic Press, New York - London 1964.

/1/ Garsia,A.M.: A simple Proof of E.Hopf's Maximal Ergodic Theorem. J.Math.Mech. 14 (1956), 381-382.

/1/ Good,I.J.: The Fractional Dimensional Theory of Continued Fractions, Proc. Cambridge 37 (1941), 199-228.

/1/ Gordin ,M.I.: Exponentially Fast Mixing.Dokl.Akad.Nauk SSSR 196 (1971), 1255-1258 = Soviet Math.Dokl. 12 (1971), 331-335.

/1/ Hall,M.Jr.: The Markoff Spectrum. Acta Arithm. 18 (1971), 387-399.

/1/ Halmos,P.R.: Lectures on Ergodic Theory. Tokyo 1953.

/2/ Halmos,P.R.: Measure Theory. New York 1968.

/1/ Hirst,K.E.: A Problem in the Fractional Dimension Theory of Continued Fractions. Quart.J.Math.Oxford Ser.(2) 21 (197o), 29-35.

/1/ Ibragimov,I.A.: A Metric Theorem in the Theory of Continued Fractions (in Russian), Vestnik Leningrad Univ. 1 (1961), 13-24.

/1/ Ibragimov,A.I. - Linnik,Yu.V.: Independent and Stationary Sequences of Random Variables, Wolters-Noordhoff, Groningen 1971.

/1/ Jacobs,K.: Lecture Notes on Ergodic Theory. Aarhus University, 1962/63.

/2/ Jacobs,K.: Neuere Methoden und Ergebnisse der Ergodentheorie Springer, Berlin 196o.

/1/ Jakubec,K.: Über Maße, die durch die Entwicklung der reellen Zahlen in Lürothsche Reihen induziert werden. Dissertation Universität Wien 1971.

/2/ Jakubec,K.: Lürothsche Reihen und singuläre Maße. Monatshefte für Mathematik.

/3/ Jakubec,K.: Induzierte Maße bei Lürothschen Reihen. Monatshefte für Mathematik.

/1/ Jarnik,V.: Zur metrischen Theorie der diophantischen Approximationen. Prace mat.fiz. 36 (1928), 91-1o6.

/1/ Kahane,J.-P. - Salem,R.: Ensembles parfaits et séries trigonométriques. Hermann,Paris 1963.

/1/ Khintchine,A.Ya.: Continued Fractions. Noordhoff,Groningen 1963.

/1/ Kinney,J.R. - Pitcher,T.S.: The Dimension of Some Sets Defined in Terms of f-Expansions. Zur Wahrscheinlichkeitstheorie verw. Geb. 4 (1966), 293-315.

/1/ Koksma,J.F.: Diophantische Approximationen. Berlin 1936.

/1/ Lévy,P.: Théorie de l'addition des variables aléatoires.Paris 1937.

/1/ Meijer,R.: Een ander bewijs voor de stelling van Schweiger over het Jacobialgorithme. Universiteit van Amsterdam 1969.

/1/ Munroe,M.E.: Introduction to Measure and Integration.Addison - Wesley,Reading 1953.

/1/ Paley,R.E.A.C. - Ursell,H.D.: Continued Fractions in Several Dimensions. Proc. Cambridge Phil. Soc. 26 (193o), 127-144.

/1/ Parry,W.: Entropy and Generators in Ergodic Theory. W.A. Benjamin, Inc. 1969.

/1/ Paysant Le Roux,R.- Dubois,E.: Algorithme de Jacobi-Perron dans un corps de séries formelles. C.R. Acad.Sci. Paris Sér. A-B 272 (1971), A 564 - A 566.

/1/ Perron,O.: Grundlagen für eine Theorie des Jacobischen Kettenbruchalgorithmus. Math.Ann. 64 (19o7), 1-76.

/1/ Philipp,W.: Mixing Sequences of Random Variables and Probabilistic Number Theory. Memoirs of the Amer.Math.Soc. 114

/2/ Philipp,W.: Some Metrical Theorems in Number Theory. Pacific J.Math. 2o (1967), 1o9-127.

/1/ Postnikov,A.G.: Ergodic Problems in the Theory of Congruences and of Diophantine Approximations. Proc.Steklov Institute Math. 82 (1966).

/1/ Rényi,A.: Representations for Real Numbers and Their Ergodic Properties. Acta Math.Acad.Sci.Hung. 8 (1957), 477-493.

/1/ Reznik,M.Kh.: The Law of the Iterated Logarithm for Some Classes of Stationary Processes (in Russian). Teoriya veroyatn. i ee primen. 13 (1968), 6o6-621.

/1/ Rogers,C.A.: Hausdorff Measures. Cambridge University Press 197o.

/1/ Ruban,A.A.: The Perron Algorithm for p-Adic numbers and Some of its Ergodic Properties. Dokl.Akad.Nauk SSSR 2o4 (1972),45-48 = Soviet Math.Dokl. 13 (1972), 6o6-6o9.

/1/ Šalât,T.: Zur metrischen Theorie der Lürothschen Entwicklungen der reellen Zahlen. Czech.Math.J. 18 (1968), 498-521.

/1/ Schmidt,W.M.: Flächenapproximation beim Jacobialgorithmus. Math. Ann. 136 (1958), 365-374.

/2/ Schmidt,W.M.: Metrical Theorems on Fractional Parts of Sequences. Trans.Amer.Math.Soc. 11o (1964), 493-518.

/1/ Schweiger,F.: Eine Bemerkung zu einer Arbeit von S.D.Chatterji. Mat.casopis 19 (1969), 89-91.

/2/ Schweiger,F.: Ergodische Theorie des Jacobischen Algorithmus. Acta Arith. 11 (1966), 451-46o.

/3/ Schweiger,F.: Existenz eines invarianten Maßes beim Jacobischen Algorithmus. Acta Arith. 12 (1967), 163-168.

/4/ Schweiger,F.: Geometrische und elementare metrische Sätze über den Jacobischen Algorithmus. Sitzungsber.Österr.Akad.Wiss., Math.-nat.Kl.Abt.II, 173 (1964), 59-92.

/5/ Schweiger,F.: Induzierte Maße und Jacobischer Algorithmus. Acta Arith. 13 (1968), 419-422.

/6/ Schweiger,F.: Metrische Sätze über den Jacobischen Algorithmus. Monatsh.Math. 69 (1965), 243-255.

/7/ Schweiger,F.: Metrische Theorie einer Klasse zahlentheoretischer Transformationen. Acta Arith. 15 (1968),1-18.

Corrigendum,Acta Arith. 16 (1969), 217-219.

/8/ Schweiger,F.: Metrische Theorie einer Klasse zahlentheoretischer Transformationen II: Hausdorffdimensionen spezieller Punktmengen. Sitzungsber. Österr.Akad.Wiss., Math.-nat.Kl.,Abt.II 177 (1969), 31-5o.

/9/ Schweiger,F.: Mischungseigenschaften und Entropie beim Jacobischen Algorithmus. J.Reine Angew.Math. 229 (1968), 5o-56.

/1o/ Schweiger,F.: Ein Kuzminscher Satz über den Jacobischen Algorithmus. J.Reine Angew.Math. 232 (1968), 35-4o.

/11/ Schweiger,F.: Abschätzung der Hausdorffdimension für Mengen mit vorgeschriebenen Häufigkeiten der Ziffern. Monatsh.Math. 76 (1972), 138-142.

/12/ Schweiger,F.: Normalität bezüglich zahlentheoretischer Transformationen. J.Number Theory 1 (1969), 39o-397.

/1/ Schweiger,F. - Stradner,W.: Die Billingsleydimension von Mengen mit vorgeschriebenen Ziffern II. Sitzungsber.Österr.Akad.Wiss., Math.-nat.Kl.Abt.II (to appear)

/2/ Schweiger,F. - Stradner,W.: Die Billingsleydimension von Mengen mit vorgeschriebenen Ziffern. Sitzungsber.Österr.Akad.Wiss.,Math.-nat.Kl. Abt.II 18o (1971), 95-1o9

/1/ Schweiger,F. - Watermann,M.S.: Some Remarks on Kuzmin's Theorem for f-Expansions. J.Number Theory (to appear)

/1/ Smorodinsky,M.: Singular Measures and Hausdorff Measures. Israel J.Math. 7 (1969), 2o3-2o6.

/1/ Szüsz,P.: Über einen Kuzminschen Satz. Acta Math.Acad.Sci.Hung.12 (1961), 447-453.

/2/ Szüsz,P.: On Kuzmin's Theorem II. Duke Math.J. 35 (1968),535-54o.

/1/ Ville,J.: Etude critique de la notion de collectif. Paris 1939.

/1/ Vinh-Hien,T.: The Central Limit Theorem for Stationary Processes Generated by Number Theoretic Endomorphisms (in Russian), Vestnik Moskov Univ.Ser.I Mat Meh. 5 (1963), 28-34.

/1/ Volkmann,B.: Über Hausdorffsche Dimensionen von Mengen, die durch Zifferneigenschaften charakterisiert sind II. Math.Z. 59(1953/54), 247-254.

/1/ Waterman ,M.S. - Beyer,W.A.: Ergodic Computations with Continued Fractions and Jacobi's Algorithm. Numer.Math. 19 (1972),195-2o5.

/1/ Waterman ,M.S.: A Kuzmin Theorem for a Class of Number Theoretic Endomorphisms. Acta Arithm. 19 (1971), 31-41.

/2/ Waterman ,M.S.: Some Ergodic Properties of Multidimensional F-Expansions. Z.Wahrscheinlichkeitstheorie verw. Geb. 16(197o), 77-1o3.

/1/ Wegmann,H.: Über den Dimensionsbegriff in Wahrscheinlichkeitsräumen. Z.Wahrscheinlichkeitstheorie verw.Geb. 9 (1968),216-231.

/1/ Yosida,K.: Functional Analysis. Springer,Berlin 1966.

Additional References

/1/ Paysant-Leroux, R. - Dubois, E.: Étude métrique de l'algorithme de Jacobi-Perron dans un corps de séries formelles. C.R.Acad.Sciences Sér. A 275 (1972), 683-686.

/1/ Schweiger, F.: Volumsapproximation beim Jacobialgorithmus. E. in Math.Annalen.

/2/ Schweiger, F.: Volumsapproximation beim Jacobialgorithmus II. E.in Acta Arithmetica.

cture Notes in Mathematics

rehensive leaflet on request

46: A. B. Altman and S. Kleiman, Introduction to Grothendieck y Theory. II, 192 pages. 1970. DM 18,–

47: D. E. Dobbs, Cech Cohomological Dimensions for Comve Rings. VI, 176 pages. 1970. DM 16,–

48: R. Azencott, Espaces de Poisson des Groupes Localement acts. IX, 141 pages. 1970. DM 16,–

49: R. G. Swan and E. G. Evans, K-Theory of Finite Groups and s. IV, 237 pages. 1970. DM 20,–

50: Heyer, Dualität lokalkompakter Gruppen. XIII, 372 Seiten. DM 20,–

51: M. Demazure et A. Grothendieck, Schémas en Groupes I. 3). XV, 562 pages. 1970. DM 24,–

52: M. Demazure et A. Grothendieck, Schémas en Groupes II. 3). IX, 654 pages. 1970. DM 24,–

53: M. Demazure et A. Grothendieck, Schémas en Groupes III. 3). VIII, 529 pages. 1970. DM 24,–

54: A. Lascoux et M. Berger, Variétés Kähleriennes Compactes. pages. 1970. DM 16,–

55: Several Complex Variables I, Maryland 1970. Edited by vath. IV, 214 pages. 1970. DM 18,–

56: R. Hartshorne, Ample Subvarieties of Algebraic Varieties. 56 pages. 1970. DM 20,–

7: T. tom Dieck, K. H. Kamps und D. Puppe, Homotopietheorie. Seiten. 1970. DM 20,–

8: T. G. Ostrom, Finite Translation Planes. IV, 112 pages. 1970. –

9: R. Ansorge und R. Hass. Konvergenz von Differenzenver für lineare und nichtlineare Anfangswertaufgaben. VIII, 145 1970. DM 16,–

0: L. Sucheston, Constributions to Ergodic Theory and Proba- VII, 277 pages. 1970. DM 20,–

61: J. Stasheff, H-Spaces from a Homotopy Point of View. pages. 1970. DM 16,–

2: Harish-Chandra and van Dijk, Harmonic Analysis on Reducadic Groups. IV, 125 pages. 1970. DM 16,–

63: P. Deligne, Equations Différentielles à Points Singuliers ers. III, 133 pages. 1970. DM 16,–

4: J. P. Ferrier, Seminaire sur les Algebres Completes. II, 69 pa- 70. DM 16,–

5: J. M. Cohen, Stable Homotopy. V, 194 pages. 1970. DM 16,–

6: A. J. Silberger, PGL_2 over the p-adics: its Representations, cal Functions, and Fourier Analysis. VII, 202 pages. 1970. –

7: Lavrentiev, Romanov and Vasiliev, Multidimensional Inverse ms for Differential Equations. V, 59 pages. 1970. DM 16,–

8: F. P. Peterson, The Steenrod Algebra and its Applications: erence to Celebrate N. E. Steenrod's Sixtieth Birthday. VII, ges. 1970. DM 22,–

: M. Raynaud, Anneaux Locaux Henséliens. V, 129 pages. 1970. –

0: Lectures in Modern Analysis and Applications III. Edited by am. VI, 213 pages. 1970. DM 18,–

Set-Valued Mappings, Selections and Topological Properties dited by W. M. Fleischman. X, 110 pages. 1970. DM 16,–

2: Y.-T. Siu and G. Trautmann, Gap-Sheaves and Extension rent Analytic Subsheaves. V, 172 pages. 1971. DM 16,–

3: J. N. Mordeson and B. Vinograde, Structure of Arbitrary Inseparable Extension Fields. IV, 138 pages. 1970. DM 16,–

4: B. Iversen, Linear Determinants with Applications to the Scheme of a Family of Algebraic Curves. VI, 69 pages. 1970. –

: M. Brelot, On Topologies and Boundaries in Potential Theory. pages. 1971. DM 18,–

6: H. Popp, Fundamentalgruppen algebraischer Mannigfaltig- IV, 154 Seiten. 1970. DM 16,–

7: J. Lambek, Torsion Theories, Additive Semantics and Rings tients. VI, 94 pages. 1971. DM 16,–

Vol. 178: Th. Bröcker und T. tom Dieck, Kobordismentheorie. XVI, 191 Seiten. 1970. DM 18,–

Vol. 179: Seminaire Bourbaki – vol. 1968/69. Exposés 347-363. IV. 295 pages. 1971. DM 22,–

Vol. 180: Seminaire Bourbaki – vol. 1969/70. Exposés 364-381. IV. 310 pages. 1971. DM 22,–

Vol. 181: F. DeMeyer and E. Ingraham, Separable Algebras over Commutative Rings. V, 157 pages. 1971. DM 16,–

Vol. 182: L. D. Baumert. Cyclic Difference Sets. VI, 166 pages. 1971. DM 16,–

Vol. 183: Analytic Theory of Differential Equations. Edited by P. F. Hsieh and A. W. J. Stoddart. VI, 225 pages. 1971. DM 20,–

Vol. 184: Symposium on Several Complex Variables, Park City, Utah, 1970. Edited by R. M. Brooks. V, 234 pages. 1971. DM 20,–

Vol. 185: Several Complex Variables II, Maryland 1970. Edited by J. Horváth. III, 287 pages. 1971. DM 24,–

Vol. 186: Recent Trends in Graph Theory. Edited by M. Capobianco/ J. B. Frechen/M. Krolik. VI, 219 pages. 1971. DM 18,–

Vol. 187: H. S. Shapiro, Topics in Approximation Theory. VIII, 275 pages. 1971. DM 22,–

Vol. 188: Symposium on Semantics of Algorithmic Languages. Edited by E. Engeler. VI, 372 pages. 1971. DM 26,–

Vol. 189: A. Weil, Dirichlet Series and Automorphic Forms. V, 164 pages. 1971. DM 16,–

Vol. 190: Martingales. A Report on a Meeting at Oberwolfach, May 17-23, 1970. Edited by H. Dinges. V, 75 pages. 1971. DM 16,–

Vol. 191: Séminaire de Probabilités V. Edited by P. A. Meyer. IV, 372 pages. 1971. DM 26,–

Vol. 192: Proceedings of Liverpool Singularities – Symposium I. Edited by C. T. C. Wall. V, 319 pages. 1971. DM 24,–

Vol. 193: Symposium on the Theory of Numerical Analysis. Edited by J. Ll. Morris. VI, 152 pages. 1971. DM 16,–

Vol. 194: M. Berger, P. Gauduchon et E. Mazet. Le Spectre d'une Variété Riemannienne. VII, 251 pages. 1971. DM 22,–

Vol. 195: Reports of the Midwest Category Seminar V. Edited by J.W. Gray and S. Mac Lane.III, 255 pages. 1971. DM 22,–

Vol. 196: H-spaces – Neuchâtel (Suisse)- Août 1970. Edited by F. Sigrist, V, 156 pages. 1971. DM 16,–

Vol. 197: Manifolds – Amsterdam 1970. Edited by N. H. Kuiper. V, 231 pages. 1971. DM 20,–

Vol. 198: M. Hervé, Analytic and Plurisubharmonic Functions in Finite and Infinite Dimensional Spaces. VI, 90 pages. 1971. DM 16,–

Vol. 199: Ch. J. Mozzochi, On the Pointwise Convergence of Fourier Series. VII, 87 pages. 1971. DM 16,–

Vol. 200: U. Neri, Singular Integrals. VII, 272 pages. 1971. DM 22,–

Vol. 201: J. H. van Lint, Coding Theory. VII, 136 pages. 1971. DM 16,–

Vol. 202: J. Benedetto, Harmonic Analysis on Totally Disconnected Sets. VIII, 261 pages. 1971. DM 22,–

Vol. 203: D. Knutson, Algebraic Spaces. VI, 261 pages. 1971. DM 22,–

Vol. 204: A. Zygmund, Intégrales Singulières. IV, 53 pages. 1971. DM 16,–

Vol. 205: Séminaire Pierre Lelong (Analyse) Année 1970. VI, 243 pages. 1971. DM 20,–

Vol. 206: Symposium on Differential Equations and Dynamical Systems. Edited by D. Chillingworth. XI, 173 pages. 1971. DM 16,–

Vol. 207: L. Bernstein, The Jacobi-Perron Algorithm – Its Theory and Application. IV, 161 pages. 1971. DM 16,–

Vol. 208: A. Grothendieck and J. P. Murre, The Tame Fundamental Group of a Formal Neighbourhood of a Divisor with Normal Crossings on a Scheme. VIII, 133 pages. 1971. DM 16,–

Vol. 209: Proceedings of Liverpool Singularities Symposium II. Edited by C. T. C. Wall. V, 280 pages. 1971. DM 22,–

Vol. 210: M. Eichler, Projective Varieties and Modular Forms. III, 118 pages. 1971. DM 16,–

Vol. 211: Théorie des Matroïdes. Edité par C. P. Bruter. III, 108 pages. 1971. DM 16,–

Please turn over

Vol. 212: B. Scarpellini, Proof Theory and Intuitionistic Systems. VII, 291 pages. 1971. DM 24,-

Vol. 213: H. Hogbe-Nlend, Théorie des Bornologies et Applications. V, 168 pages. 1971. DM 18,-

Vol. 214: M. Smorodinsky, Ergodic Theory, Entropy. V, 64 pages. 1971. DM 16,-

Vol. 215: P. Antonelli, D. Burghelea and P. J. Kahn, The Concordance-Homotopy Groups of Geometric Automorphism Groups. X, 140 pages. 1971. DM 16,-

Vol. 216: H. Maaß, Siegel's Modular Forms and Dirichlet Series. VII, 328 pages. 1971. DM 20,-

Vol. 217: T. J. Jech, Lectures in Set Theory with Particular Emphasis on the Method of Forcing. V, 137 pages. 1971. DM 16,-

Vol. 218: C. P. Schnorr, Zufälligkeit und Wahrscheinlichkeit. IV, 212 Seiten 1971. DM 20,-

Vol. 219: N. L. Alling and N. Greenleaf, Foundations of the Theory of Klein Surfaces. IX, 117 pages. 1971. DM 16,-

Vol. 220: W. A. Coppel, Disconjugacy. V, 148 pages. 1971. DM 16,-

Vol. 221: P. Gabriel und F. Ulmer, Lokal präsentierbare Kategorien. V, 200 Seiten. 1971. DM 18,-

Vol. 222: C. Meghea, Compactification des Espaces Harmoniques. III, 108 pages. 1971. DM 16,-

Vol. 223: U. Felgner, Models of ZF-Set Theory. VI, 173 pages. 1971. DM 16,-

Vol. 224: Revêtements Etales et Groupe Fondamental. (SGA 1). Dirigé par A. Grothendieck XXII, 447 pages. 1971. DM 30,-

Vol. 225: Théorie des Intersections et Théorème de Riemann-Roch. (SGA 6). Dirigé par P. Berthelot, A. Grothendieck et L. Illusie. XII, 700 pages. 1971. DM 40,-

Vol. 226: Seminar on Potential Theory, II. Edited by H. Bauer. IV, 170 pages. 1971. DM 18,-

Vol. 227: H. L. Montgomery, Topics in Multiplicative Number Theory. IX, 178 pages. 1971. DM 18,-

Vol. 228: Conference on Applications of Numerical Analysis. Edited by J. Ll. Morris. X, 358 pages. 1971. DM 26,-

Vol. 229: J. Väisälä, Lectures on n-Dimensional Quasiconformal Mappings. XIV, 144 pages. 1971. DM 16,-

Vol. 230: L. Waelbroeck, Topological Vector Spaces and Algebras. VII, 158 pages. 1971. DM 16,-

Vol. 231: H. Reiter, L¹-Algebras and Segal Algebras. XI, 113 pages. 1971. DM 16,-

Vol. 232: T. H. Ganelius, Tauberian Remainder Theorems. VI, 75 pages. 1971. DM 16,-

Vol. 233: C. P. Tsokos and W. J. Padgett, Random Integral Equations with Applications to Stochastic Systems. VII, 174 pages. 1971. DM 18,-

Vol. 234: A. Andreotti and W. Stoll, Analytic and Algebraic Dependence of Meromorphic Functions. III, 390 pages. 1971. DM 26,-

Vol. 235: Global Differentiable Dynamics. Edited by O. Hájek, A. J. Lohwater, and R. McCann. X, 140 pages. 1971. DM 16,-

Vol. 236: M. Barr, P. A. Grillet, and D. H. van Osdol. Exact Categories and Categories of Sheaves. VII, 239 pages. 1971. DM 20,-

Vol. 237: B. Stenström. Rings and Modules of Quotients. VII, 136 pages. 1971. DM 16,-

Vol. 238: Der kanonische Modul eines Cohen-Macaulay-Rings. Herausgegeben von Jürgen Herzog und Ernst Kunz. VI, 103 Seiten. 1971. DM 16,-

Vol. 239: L. Illusie, Complexe Cotangent et Déformations I. XV, 355 pages. 1971. DM 26,-

Vol. 240: A. Kerber, Representations of Permutation Groups I. VII, 192 pages. 1971. DM 18,-

Vol. 241: S. Kaneyuki, Homogeneous Bounded Domains and Siegel Domains. V, 89 pages. 1971. DM 16,-

Vol. 242: R. R. Coifman et G. Weiss, Analyse Harmonique Non-Commutative sur Certains Espaces. V, 160 pages. 1971. DM 16,-

Vol. 243: Japan-United States Seminar on Ordinary Differential and Functional Equations. Edited by M. Urabe. VIII, 332 pages. 1971. DM 26,-

Vol. 244: Séminaire Bourbaki - vol. 1970/71. Exposés 382-399. IV, 356 pages. 1971. DM 26,-

Vol. 245: D. E. Cohen, Groups of Cohomological Dimension One. V, 99 pages. 1972. DM 16,-

Vol. 246: Lectures on Rings and Modules. Tulane University and Operator Theory Year, 1970-1971. Volume I. X, 661 pages. DM 40,-

Vol. 247: Lectures on Operator Algebras. Tulane University Ring Operator Theory Year, 1970-1971. Volume II. XI, 786 pages. DM 40,-

Vol. 248: Lectures on the Applications of Sheaves to Ring Th Tulane University Ring and Operator Theory Year, 1970-1971. ume III. VIII, 315 pages. 1971. DM 26,-

Vol. 249: Symposium on Algebraic Topology. Edited by P. J. H VII, 111 pages. 1971. DM 16,-

Vol. 250: B. Jónsson, Topics in Universal Algebra. VI, 220 pa 1972. DM 20,-

Vol. 251: The Theory of Arithmetic Functions. Edited by A. A. and D. L. Goldsmith VI, 287 pages. 1972. DM 24,-

Vol. 252: D. A. Stone, Stratified Polyhedra. IX, 193 pages. DM 18,-

Vol. 253: V. Komkov, Optimal Control Theory for the Dampin Vibrations of Simple Elastic Systems. V, 240 pages. 1972. DM

Vol. 254: C. U. Jensen, Les Foncteurs Dérivés de lim et leurs plications en Théorie des Modules. V, 103 pages. 1972. DM

Vol. 255: Conference in Mathematical Logic - London '70. Edite W. Hodges. VIII, 351 pages. 1972. DM 26,-

Vol. 256: C. A. Berenstein and M. A. Dostal, Analytically Uni Spaces and their Applications to Convolution Equations. VII, pages. 1972. DM 16,-

Vol. 257: R. B. Holmes, A Course on Optimization and Best proximation. VIII, 233 pages. 1972. DM 20,-

Vol. 258: Séminaire de Probabilités VI. Edited by P. A. Meye 253 pages. 1972. DM 22,-

Vol. 259: N. Moulis, Structures de Fredholm sur les Variétés bertiennes. V, 123 pages. 1972. DM 16,-

Vol. 260: R. Godement and H. Jacquet, Zeta Functions of Si Algebras. IX, 188 pages. 1972. DM 18,-

Vol. 261: A. Guichardet, Symmetric Hilbert Spaces and Related pics. V, 197 pages. 1972. DM 18,-

Vol. 262: H. G. Zimmer, Computational Problems, Methods, Results in Algebraic Number Theory. V, 103 pages. 1972. DM

Vol. 263: T. Parthasarathy, Selection Theorems and their Applicat VII, 101 pages. 1972. DM 16,-

Vol. 264: W. Messing, The Crystals Associated to Barsotti Groups: with Applications to Abelian Schemes. III, 190 pages. DM 18,-

Vol. 265: N. Saavedra Rivano, Catégories Tannakiennes. II, pages. 1972. DM 26,-

Vol. 266: Conference on Harmonic Analysis. Edited by D. G and R. L. Lipsman. VI, 323 pages. 1972. DM 24,-

Vol. 267: Numerische Lösung nichtlinearer partieller Differential Integro-Differentialgleichungen. Herausgegeben von R. Ansorge W. Törnig. VI, 339 Seiten. 1972. DM 26,-

Vol. 268: C. G. Simader, On Dirichlet's Boundary Value Problem. I pages. 1972. DM 20,-

Vol. 269: Théorie des Topos et Cohomologie Etale des Sch (SGA 4). Dirigé par M. Artin, A. Grothendieck et J. L. Verdier. 525 pages. 1972. DM 50,-

Vol. 270: Théorie des Topos et Cohomologie Etle des Sch Tome 2. (SGA 4). Dirige par M. Artin, A. Grothendieck et J. L. Ve V, 418 pages. 1972. DM 50,-

Vol. 271: J. P. May, The Geometry of Iterated Loop Spaces. IX pages. 1972. DM 18,-

Vol. 272: K. R. Parthasarathy and K. Schmidt, Positive Definite nels, Continuous Tensor Products, and Central Limit Theorem Probability Theory. VI, 107 pages. 1972. DM 16,-

Vol. 273: U. Seip, Kompakt erzeugte Vektorräume und Analys 119 Seiten. 1972. DM 16,-

Vol. 274: Toposes, Algebraic Geometry and Logic. Edited by. Lawvere. VI, 189 pages. 1972. DM 18,-

Vol. 275: Séminaire Pierre Lelong (Analyse) Année 1970-197 181 pages. 1972. DM 18,-

Vol. 276: A. Borel, Représentations de Groupes Localement pacts. V, 98 pages. 1972. DM 16,-

Vol. 277: Séminaire Banach. Edité par C. Houzel. VII, 229 p 1972. DM 20,-